THE
MIRACLE MYTH

THE

MIRACLE MYTH

WHY BELIEF IN THE RESURRECTION AND
THE SUPERNATURAL IS UNJUSTIFIED

LARRY SHAPIRO

COLUMBIA UNIVERSITY PRESS   *New York*

Columbia University Press

Publishers Since 1893

New York   Chichester, West Sussex

cup.columbia.edu

Library of Congress Cataloging-in-Publication Data

Names: Shapiro, Lawrence A., author.

Title: The miracle myth : why belief in the Resurrection and the supernatural

is unjustified / Larry Shapiro.

Description: New York City : Columbia University Press, 2016. | Includes

bibliographical references and index.

Identifiers: LCCN 2016002183 | ISBN 9780231178402 (cloth : alk. paper) |

ISBN 9780231542142 (e-book)

Subjects: LCSH: Supernatural. | Miracles. | Belief and doubt. | Credulity. | Psychology.

Classification: LCC BL100 .S43 2016 | DDC 202/.117—dc23

LC record available at https://lccn.loc.gov/2016002183

Columbia University Press books are printed on permanent and durable acid-free paper.

Printed in the United States of America

c 10 9 8 7 6 5 4 3 2 1

COVER DESIGN: CHANG JAE LEE

COVER IMAGE: *RESURRECTION OF CHRIST* BY RAPHAEL

© ALFREDO DAGLI ORTI / THE ART ARCHIVE, CORBIS

To my parents, Martin and Ann,
who gave me the miracle of life but, more importantly, raised me right

CONTENTS

PREFACE ix

ACKNOWLEDGMENTS xvii

1. Justified and Unjustified Belief
1

2. Miracles
17

3. Justifying Belief in Supernatural Causes
29

4. Justifying Belief in Improbable Events
57

5. Evidence for Miracles
87

6. Jesus's Resurrection
109

7. Should We Care That Beliefs in Miracles Are Unjustified?
137

APPENDIX 1. WHAT IS SUPERNATURAL? 149

APPENDIX 2. SUPERNATURAL CAUSES 155

NOTES 159

FURTHER READING 163

INDEX 165

PREFACE

MANY PEOPLE BELIEVE IN MIRACLES. IN FACT, IN 2010 THE
Pew Forum on Religion asked thirty-five thousand Americans, age eigh-
teen or older, whether they agree that "miracles still occur today as in
ancient times."[1] The results show that 47 percent of participants "completely agree" and another 32 percent "mostly agree." If the sample is
representative, this means that nearly four out of five American adults
believe in miracles. Obviously, with so many people in agreement, belief
in miracles must cut across all sorts of the usual categories—liberal and
conservative, young and old, gay and straight, southern and northern. It
turns out that there is no single "type" who believes. We can't even conclude that believers are particularly religious. Further analysis of the
survey reveals that even nonreligious people—those who don't identify
with any particular religion and never attend houses of worship—believe in miracles. Even atheists and agnostics believe in miracles! Miracles

don't belong only to the very conservative religious fringe. They are accepted as fact by a huge majority of people in the United States.

What's more, the context in which the question about miracles was asked made clear that the miracles being referred to were the BIG ones. We're not talking simple coincidences, as in the time you sat on an airplane next to a stranger who turned out to be the person who would interview you for a job later that day. We're also not talking about merely unexpected occurrences, as in when the U.S. Olympic hockey team beat the Russians in the so-called Miracle on Ice. Coincidences and other unlikely events are not what come to mind when thinking about miracles "in ancient times." Seas parting on command, bodies rising from the dead, water suddenly turning into wine—*these* are the miracles that we associate with ancient times, and they make coincidences and improbabilities, surprising as they may be, seem rinky-dink by comparison.

Here's another set of questions the Pew survey asked: "Do you think there is a heaven, where people who have led good lives are eternally rewarded?" and "Do you think there is a hell, where people who have led bad lives and die without being sorry are eternally punished?" Although the focus of this book is mainly miracles, I am mentioning these questions about heaven and hell because the way the questions were answered will help me to introduce one of my main themes. Of those surveyed, 74 percent believe that heaven exists, 59 percent believe in hell. Speaking now as a trained philosopher, I find this result very puzzling. Let me explain why.

Philosophers like to argue that when people believe something, they should have a *reason* to do so. If you believe that rain is on the way, you should have a reason that makes such a belief appropriate. Maybe you see dark clouds on the horizon, or maybe you heard a forecast on the radio earlier in the day. Likewise, if you believe that your car needs to go to the shop, you again should have a reason, such as that grinding noise you hear when you put it in reverse or the warning lights on the dashboard that flash whenever you start the engine. Asking for a reason to believe something doesn't seem like a big deal. Even a *bad* reason is a reason. Maybe you think it's going to rain because your horoscope said so or because your dog is barking. These reasons may not be great ones, but they're something. They show that some effort is being made to justify your belief, even if the effort turns out not to be sufficient to that task.

I bet most people understand that they have a responsibility to give a reason when someone asks them why they believe that rain is on the way or why they believe that their car needs to be repaired. "I just believe" is hardly a satisfying answer. I think I would resent someone who convinced me to carry an umbrella to work if he had no *reason* to believe that it would rain, and I *know* I would resent someone who persuaded me to spend money to fix a car that she had no *reason* to think was broken.

But now let's consider the survey responses to the questions about heaven and hell. The data show that more people believe in heaven than people who believe in hell (specifying exactly how many more is difficult because some respondents refused to answer the questions or checked "I don't know"). I find this discrepancy odd because the existence of heaven seems no more or less likely than the existence of hell. Whatever *reason* to believe that heaven exists—it's a reward God offers to those who have lived a worthy life—seems to suggest quite naturally a reason to believe that hell also does—as punishment for those who do not live the life God commands. Because reasons for believing in heaven seem no better or worse than reasons to believe in hell, why do so many Americans accept the idea of one but not the other?

I do have an explanation for the survey results, but I confess that it's not one that makes me happy. In fact, it's an explanation that curls most philosophers' nose hairs. Here it is: more people believe in heaven than in hell not because they have reasons to believe in one and not the other but because they *want it to be true* that there's a heaven and *don't want it to be true* that there's a hell. In other words, the people who believe in heaven but not hell have abandoned reason and instead hang their belief on nothing more than hope—hope that because the idea of heaven is so nice, it must exist, and hope that because the idea of hell is so horrible, it must not exist. I think this dependence on hope, believing what you really want to be true and not believing what you really want to be false, explains the survey results. Put more simply, in a way that's not at all surprising, people believe what they want to believe.

What's wrong with believing what you want to believe? Why is having a reason to believe something better than simply believing it because you *want* it to be true? These are questions that philosophers don't usually try to answer because, as professional reasoners, most of them just assume that everyone shares a devotion to having reasons. But we can say a few things about why it's wrong to believe something simply because you really want it to be true.

For one thing, wanting something to be true doesn't make it true. Here's an easy way to demonstrate this to yourself. Imagine something that you really love, and then wish for it. When I engage in this practice, chocolate usually springs to mind before anything else. I am now imagining a dark-chocolate truffle, covered in cocoa powder. I would so much like it to be true that a dark-chocolate truffle will now appear on the desktop before me. The more I think about it, the more and more I want it to be true. I'm going to close my eyes now, and I intensely hope that when I reopen them in a second, I will see a dark-chocolate truffle sitting in a small piece of glitzy foil, inches from my finger tips. . . .

Nothing. As despondent as this exercise makes me—and for that reason I do not engage in it very often—I do take comfort that it shows the philosopher's love for reasons to be, well, reasonable. It shows that merely wanting something to be true doesn't make it true. If I am to believe that a chocolate truffle will appear before me, I need a better reason than simply wanting one to appear. I need to have *reasons* for believing. Wanting isn't a reason. Wanting, hoping, and wishing have nothing to do with truth. But if someone told me that she would be delivering truffles to my study later today, or if I knew that today was my birthday and my children traditionally rain truffles upon me on this greatest of days, then my belief that I'll soon be eating a truffle has some support behind it. My belief now has a better chance of being true than if it were to rest on mere wishes.

Reasons for believing something—or, as we'll see more clearly in a moment, *good* reasons for believing something—bear on truth in a way that wanting, hoping, and wishing cannot. If I have a good reason to believe that my team the Green Bay Packers will win the Super Bowl, that's just another way of saying that the sentence "The Green Bay Packers will win the Super Bowl" is more likely to be true than if there were no reason. Really, really wanting the Packers to win but having no good reason to think that they will tells you nothing about how likely it is that the sentence is true. Having a good reason to believe that you'll earn a million dollars means that the sentence "I will earn a million dollars" has a greater probability of being true than if you have no reason. Merely wanting it to be true that you'll earn a million dollars tells you nothing about whether you ever will. Hope and wish all you want, unless you have a *good reason* to believe something, the truth of what you believe can't be counted on (I discuss what makes a reason good later in this preface and throughout the book).

So if I'm right that more people believe in heaven than in hell just because they hope that there's one but not the other, then their belief says nothing about the truth of whether there is a heaven or a hell. They're on no firmer ground in believing that there's a heaven than I am in believing that a chocolate truffle will appear before me just because I really want one.

As I said, I'm a philosopher. As corny as this sounds, I'm dedicated to discovering the truth. This is why I'm so concerned with having reasons. If I'm going to believe in heaven, I need a reason. You wouldn't step aboard an airplane if you had a *reason* to think it was unsafe, even if you really *wished* it were safe. Reasons trump wishes. We know that. Hoping won't get me chocolate truffles, it won't create a heaven, and it doesn't justify belief in miracles either.

Then what about the 80 percent of Americans who believe in miracles? No doubt some of these people believe in miracles for no reason at all. They just really want it to be true that miracles have occurred. They *hope* that there have been miracles. To these people, I have nothing to say. I have no patience for hopers and wishers. Hoping won't make it safe to fly on a broken airplane. That's just not how the world works, and when you think about it, that might not be such a bad thing. Just imagine if you could make something true by hoping it into being. Those who wished harm on others would cause people to suffer; those who hoped for the end of the world would ruin things for the rest of us. But maybe not. What happens when I wish for one thing and you wish for the opposite? If hope were enough to make something true, how do we explain that one team beats another when players on both teams hope for victory? How do we explain why poor people remain poor or why sick people stay sick? The more you think about it, the sillier the idea becomes that just by wanting something to be true, you can make it true.

Hoping won't get you a miracle. If you want your belief in miracles to have a decent chance of actually being true, you need reasons. This book is written for those of you who think you have good reasons to believe in miracles. Maybe you believe in miracles because you trust biblical accounts of such things. Could all the stuff in Matthew, Mark, Luke, and John really be just made up? Or maybe you believe in miracles because you have read accounts of crying statues or of terminally sick cancer patients awaking one day with no signs of illness. Surely, you might think, too many reports like these exist to doubt them all. Maybe you have witnessed with your own eyes what you think is a miracle or know someone who has. I grant that all of these instances provide reasons to believe in miracles—reasons that, unlike mere hopes and wishes, strive to say something about the truth of miracles.

But, of course, there are reasons, and then there are reasons. Not all reasons are good reasons. If I believe that my Green Bay Packers will win the Super Bowl this year because I have studied the players' statistics, researched the opposition, weighed the strengths and weaknesses of the defense, examined the medical records of the offensive line, and so on, then I'm onto something. My belief in my team is in some sense *justified*. But if I believe the Packers will win because of something I saw in my tea leaves or because I think that the ghost of Vince Lombardi will intervene on the team's behalf or because of the way the planets are aligned, then, even though I have reasons— I'm not believing on just the basis of hope—I don't have the *kind* of reasons that tell me anything about whether the Packers *really* have a chance to go all the way this year. For that, I'll need good reasons. Bad reasons should be commended for trying, but in the end they're no better at tracking the truth than are hopes and wishes.

My goal in this book is to convince you that no one has ever had or currently has good reasons for believing in miracles. The reasons people give for believing in miracles, such as the second kind of reasons I mentioned for believing that the Packers will win the Super Bowl, are bad. They don't tell you anything about whether miracles have actually ever occurred. This means that you shouldn't believe in miracles. Or, if you want to continue to believe in miracles, you should do so with the recognition that your belief is based on bad reasons or something other than reason—for instance, hope or, what amounts to pretty much the same thing, faith. But I have said enough about what I think of that.

One more word before I start. Although I do think that no one should believe in miracles, my conclusion isn't all bad news for those 80 percent of Americans who do believe, and it's not all good news for those 20 percent who don't. My focus is on *belief*—on whether anyone's *belief* in miracles can be justified. But this target is completely independent of another issue—the issue of whether *in fact* there have been miracles. As I explain in chapter 1, you can be justified in believing in things that *do not* really exist, just as you may *not* be justified in believing in things that do. Are we justified in believing in miracles? No. We are not. Have there ever been any miracles? Personally, I doubt it, but I confess that I don't know how to argue for that conclusion.

ACKNOWLEDGMENTS

MANY PEOPLE HAVE DISCUSSED THE MATERIAL IN THIS BOOK with me or provided me with detailed notes on various drafts. I am grateful to all of them, but especially to my wife, Athena Skaleris, who not only read the entire manuscript but also tolerated me through the process of researching, writing, and publishing. I hope I haven't forgotten to name anyone else in the following list, but if I did, I promise to buy that person a beer when next we meet! Thanks to Michael Lynch, Steve Nadler, Ron Numbers, Sarah Paul, Russ Shafer-Landau, Alan Sidelle, David Skeel, Elliott Sober, Mike Titelbaum, and Peter Vranas. Wendy Lochner and Christine Dunbar of Columbia University Press have made the publishing process a pleasure. Annie Barva's edits were unerring. I'm very grateful to Kathryn Jorge for her help in readying the manuscript for production. Thanks also to the Institute for Research in the Humanities at the University of Wisconsin for the support it provided me.

THE
MIRACLE MYTH

I

JUSTIFIED AND UNJUSTIFIED BELIEF

LET'S START WITH A STORY THAT SHOULD PUT SOME COMMON ground under the feet of both believers in miracles and nonbelievers. In the chapters that follow, we'll return now and then to the story. This way, if you find yourself disagreeing with the direction I take, you can retrace the path from our shared starting point to figure out where we began to diverge. This strategy, I hope, will help to sharpen our differences and might even lead you to clarify in your own mind why you believe in miracles if in fact you do.

So here's the story. You're sitting at a bar, chatting amiably with the fellow who straddles the stool next to you. He seems ordinary enough in his blue windbreaker and khaki pants. Clean shaven, brown hair receding a bit, he's probably about forty-five years old. In the time it's taken you to drink your first beer, you have discovered that you and he share many of the same life experiences. Both of you are married, have teenage children, hold white-collar jobs, and like to watch cooking programs on cable TV. He's a Minnesota Vikings fan, but you try not to hold that against him. You steer clear of politics—not because you're worried that you'll get into a heated discussion but just because it's one of those conversations that ends up depressing you even when you find yourself agreeing with everything that's being said.

The more you talk, the more you like the guy. You get around to introducing yourselves by name. He's Jim. Jim likes to read crime novels, and you share a passion for Lawrence Block's Bernie Rhodenbarr series. He confides in you that he believes Lucille Ball was the most beautiful woman who's ever lived. OK, you think, that's a little weird, but there's a case to be made and, anyway, there's no accounting for taste. The afternoon stretches into evening.

"And I've got one for you," Jim says, after you've finished a joke about a guy who walks into a bar with a shoebox that contains a twelve-inch pianist. You order another beer, settling comfortably on your bar stool in anticipation of the joke to follow. "True story," he begins, in just the way you expect a funny story to start. "There's a frog that lives on a lily pad in the middle of a lake in the Karnataka region of India. It can speak, and not just in the native tongues Urdu and Hindi but also German, English, French, and Navajo." You smile, imagining the frog in conversation with a Navajo. "And people who have visited the frog," he continues, "report marvelous happenings. Some have been cured of cancer, others find amputated limbs regenerating, and still others have brought dead pets to the frog. It waves its little arms around and says some things in a language no one understands, and the pets come back to life." Jim pauses, looking up at the ceiling as if in thought. "Never ferrets, though," he mumbles. "I wonder what's up with that?" He then looks into his mug, which is nearly empty now. You stir in your chair, waiting for the punch line. Was that it? "Never ferrets?" Did you miss something, or maybe it was just a lousy joke?

"Yeah," Jim finally says, looking into his empty mug. "Someday I'm going to visit that frog." Oh, you think. Maybe that was the punch line. Ha ha. You wait for more, but your new friend seems to have finished. He catches the bartender's eye and gestures for another beer.

"Was that it?" you ask, trying not to show your annoyance. "Not much of a joke."

"Joke?" he says, raising his eyebrows. "That was no joke. Like I told you, it's a true story. That frog's as real as this bar," he says, rapping his knuckles on the polished wood in front of you. You look closely at Jim's face for signs that he might be pulling your leg. "Come on, Jim," you think. "Crack a smile or wink or do *something* that tells me you're only kidding." But he just leans forward to grab the newly replenished mug.

"You're serious?" you say weakly, hope fading that you can salvage your budding friendship. "You really think that a magic frog lives in some place called Karntaka?"

"Karnataka," Jim corrects you. "And yes. I really believe it."

"But you haven't seen it yourself," you continue.

"Nope."

"So what makes you think it really exists?" you ask, making an effort to keep your tone respectful.

"I've got friends who've seen it," he says. "Well," he hurries to add, "not actually. But friends of friends. Or maybe, to tell the truth, it's friends of friends of friends. Whatever. I've heard about the frog, and I believe what I've been told."

End of story.

Common Ground

I said that my intent in offering this story is to provide believers and non-believers in miracles with some common ground. I hope the story does not offend believers, who might take umbrage at what they see as a haughty suggestion that the miracles reported in, say, the New Testament are on par with a moderately inebriated person's tale of a magical frog. We will have ample time in the following chapters to assess this possibility.

For now, as much as we can, let's detach the story of the frog from any thoughts about miracles. I promise that we'll come to miracles soon enough. Here's what I want us to agree on: Jim really does believe in the existence of a frog in Karnataka that speaks many languages and performs incredible deeds, but that doesn't mean that he is *justified* in believing this story. This difference between believing something and being *justified* in believing something is absolutely crucial, so in the next section I explain how I understand the difference. However, I expect that even now the distinction sounds familiar to most of us. We all recognize the difference between simply believing that our favorite team will win the Super Bowl and being justified in believing it will. In the former, the belief is more like a wish, as I discussed already in the preface to this book. In the latter, we think we have evidence that actually speaks to the probability that our team will win. An unjustified belief is often like a wish—something you believe not because the available evidence convinces you it must be true but because you *want* it to be true. We will eventually be in a position to see why Jim's belief in the miraculous Karnatakan frog is more like a wish. He might want to believe that it exists, but he is not justified in believing that it does.

Furthermore, I think we should agree that whether Jim is currently justified in believing in the existence of the miraculous frog is distinct from the question about whether the frog really does exist. This assertion might seem strange or surprising at first. If the frog *does* exist, why isn't Jim justified in believing that it does? I have more to say about this distinction later, but the basic idea is that the existence of a miraculous frog is one thing; whether anyone is justified in believing in its existence is quite another. This distinction, on reflection, should also not seem so peculiar. We see it in action all the time in courtroom dramas, where the guilt or innocence of a defendant is in question. Whether the defendant is guilty is one thing; whether the jury is *justified* in believing that the defendant is guilty is quite another. The prosecutor's job is to provide the members of the jury with enough evidence to make them justified in believing in the defendant's guilt. If the prosecutor is good, she will be able to do this. In fact, she will be able to do it whether the defendant really is guilty or not. Similarly, I am claiming here that we can separate the question of whether Jim's belief in the miracle-working frog is justified from the question of whether his belief is true—that is, whether the frog really exists.

My job in the next section is to clarify and defend the remarks I have made about Jim and the frog. Of course, you might already be comfortable with the idea that Jim is free to believe in the Karnatakan frog's existence without being justified in his belief. And maybe you're prepared to agree that Jim's belief in the frog might be unjustified even if the frog actually exists. If so, reading the next section should help to illuminate our reasons for agreement. However, if you're wondering why we should agree on these points, then I look forward in the next section to convincing you that I am right.

A Little Epistemology

The philosopher's interest in knowledge is as old as philosophy itself, and that's pretty old. *Epistemology* is just a fancy word for the field of philosophy that concerns itself with questions about knowledge and justification. For example, philosophers interested in epistemology ask questions such as "Is it possible to know anything?" This question might seem like one of those silly matters that make philosophers an easy target for people who actually work for a living. Of course we can know things, you might say. Here's a list of things we know: water freezes at 32°F; whales are mammals; 2 + 3 = 5; Australia is in the Southern Hemisphere; a square has four sides of equal length.

I agree that I know all the propositions I just mentioned, and, moreover, most (but not all!) philosophers would concede the same. Very few philosophers are skeptics about knowledge. But one way to understand the philosopher's question about *what* we can know is a subtler question about what it *means* to know something. Or, put another way, the philosopher wants to know what knowledge is. We clearly know some things—such as those things mentioned earlier—but not others. For instance, we know that other planets exist, but we don't know whether there is life on other planets (although we may believe that there is). We know that there exists no largest prime number because Euclid proved it, but we don't know whether every even number greater than two can be expressed as the sum of two prime numbers (the claim that they can be is called *Goldbach's conjecture*). But what makes some beliefs count as knowledge but others not? Without an answer to this question, we are straight back to wondering whether we can know anything.

But this book isn't about knowledge; it's about justification. Justification requires less than knowledge does in the sense that you can be justified in believing a proposition even if you don't know it, but you can't know something without having justification for believing it. It takes more to know something than it does to be justified in believing it. This means that even though my focus is on whether belief in miracles can be *justified*, my conclusion has consequences for whether we can *know* that there have been miracles. Because knowledge depends on justification, if we're not justified in believing in miracles, we can't know that they have occurred.

Very important for our purposes is the distinction between *justified* belief and *unjustified* belief. Here's an example of an unjustified belief that my mother-in-law had. She thought that because she had already had four daughters, her fifth child would be a boy. This belief was unjustified because births have no influence on each other. Having had a bunch of daughters doesn't increase the probability that your next child will be a son. To think it does is to commit the notorious gambler's fallacy of thinking that you should bet red on roulette because the ball has landed on black five times in a row, and so now the odds must be better that the ball will land on red. They're not better. Same goes for coin flips. If the coin is fair, ten heads in a row tells you absolutely nothing about whether the eleventh flip is more or less likely to land tails up.

I might also point out that even though my mother-in-law was not justified in believing that her fifth child would be a son, her belief *may have* been true (but it wasn't—she gave up on having a boy after five girls). Whether a belief is true or false doesn't depend on whether it is justified but on how the world is. You can have true but unjustified beliefs, just as you can have false *justified* beliefs, as we'll soon see.

To understand the extent of our unjustified beliefs, I need to draw a further distinction between having a justification for believing something and having an *explanation* for believing something. This is a more precise way of drawing a distinction I mentioned in the preface between good reasons and bad reasons. Making this distinction is necessary because some people might deny that bad reasons are reasons at all, just as some might deny that a joke that isn't funny is really a joke. So to make everyone happy, let's use the word *reason* to refer to something that raises the chance of some belief being true or, equivalently for our purposes, to something that contributes to justification for the belief. In contrast, I use the word *explanation* to refer to something that does nothing to raise the chance that a belief is true.

My mother-in-law, then, could *explain* why she believed that her fifth child would be a boy: she had already had four daughters; a run of four daughters was very unlikely, she thought, and a run of five daughters would be even more unlikely. Therefore, the chance that her next child would be a boy was high. Although this is an *explanation* for her belief, it's not a justification.

I think this distinction between explanations and justifications is especially intuitive when talking about morality. The schoolyard bully might be able to *explain* why he has beaten up a classmate—"I didn't like his face!"—but that assertion hardly *justifies* his actions. His behavior was wrong even though he certainly could explain why he behaved as he did. This example shows that explanation and justification are distinct notions. The distinction holds in my mother-in-law's case as well. She could *explain* why she believed that her next baby would be a boy, but her explanation failed to *justify* her belief.

How do we know when reasons for believing something count as a justification and when they do not? Justification, as philosophers understand it, should raise the probability that the belief really is true. One way to make this clear is by contrasting true beliefs that are justified with true beliefs that are not. In the latter case, luck plays a role that it does not in the former. Suppose, for instance, that my mother-in-law's belief turned out to be true: her fifth child was in fact a boy. If the belief were true, its truth would not be for the explanation she provided for it. Her explanation was bogus. Rather, she would have been right about the sex of her fifth child just because her chances of having a boy were roughly 50 percent, and it just so happened that (we're supposing) a boy is what she had.

However, when we look at justified beliefs, the role of luck is minimized. Consider another case. You heft a bag of Halloween candy that you have removed from your child's bedroom. Halloween was two weeks ago, and you have decided that she's consumed enough sweets by now. But that doesn't mean that you shouldn't enjoy a piece. You reach into the bag hoping for a Reese's peanut butter cup and pull out a Heath bar. Yuck. You hate Heath bars. Tossing it into the trash, you reach into the bag again and pull out another Heath bar. This too you fling into the trash. After repeating this same operation twenty times, and with only one piece of candy left in the bag, you form the belief that the next piece of candy you remove from the bag will be a Heath bar. Sure enough, you choke back your tears as you extract yet another Heath bar from the bag. Your belief was true.

The basis for your belief that the last piece of candy in the trick-or-treat bag is a Heath bar differs dramatically from the basis of my mother-in-law's belief about the sex of her next child. Both cases, we're imagining, might involve a true belief, but whereas my mother-in-law's belief was true "by luck," your belief about the next piece of candy was not. The more Heath bars you pull from the bag, the more likely it becomes that Heath bars are the only things that remain. Why? Odds are that if your daughter had left a variety of different candies in the bag, you would have pulled out something else by the time you reach the last few pieces—a Reese's peanut butter cup, perhaps. Clearly, your daughter shares your dislike of Heath bars.

Many beliefs that people hold lack justification. Some people believe that crystals have a power to heal cancer or prevent baldness or that large hairy primates roam America's woodlands. I think these beliefs are silly, but that's not to say that people don't have explanations for them. I recently saw a television program about the search for Sasquatches. It depressed me terribly. A person reported being in a tent deep in the woods on a dark night. Something pressed against his tent, startling him awake. He suspected it was a Sasquatch. The Sasquatch "expert" interviewing him asked whether he had seen the beast, but he hadn't. The expert then said that this was really good evidence that the creature that disturbed the camper was indeed a Sasquatch because Sasquatches are rarely seen.

TABLE 1.1 Kinds of True Beliefs

	TRUE BELIEFS
JUSTIFIED	The next piece of candy is a Heath bar.
UNJUSTIFIED	The next child will be a boy.

The camper seemed satisfied with this "justification" for his belief that a Sasquatch really had been present. (Is *not* seeing a Sasquatch better evidence for believing in them than actually seeing one? I wonder.) Similarly, you might mention your horoscope to explain why you believe that you should not leave your house today, but a horoscope cannot justify your belief. Even if the belief that you shouldn't leave your house today turns out to be *true* (when you do, you slip on some ice and break your wrist), it's only luck that connects your horoscope to your misfortune.

To keep track of the relationships between the truth of a belief and its justification, let's begin to build a table. Table 1.1 illustrates the distinction between true beliefs that are justified, such as the belief you formed on the basis of pulling candy from your daughter's trick-or-treat bag, and true beliefs that are not justified. An example of the latter is my mother-in-law's belief that her fifth child will be a boy (assuming it was). This belief, even if it turned out to be true, would not be justified because the explanation offered in its support is defective: the sexes of her first four children do not provide evidence about the sex of her subsequent child.

Now let's think about expanding this table. We have seen that true beliefs can be either justified or not. The same holds of false beliefs. The possibility of false, unjustified beliefs should hardly be surprising. We have seen already the sense in which unjustified beliefs, if true, are true as a matter of luck. My mother-in-law's fifth child just happened to be (we're supposing) a boy. But it just as easily could have turned out to be a girl. If it were a girl, her belief that she was going to have a boy was both false and unjustified. Justification should do something to increase the odds of a belief being true.

But merely because justification works to increase the probability that a belief is true, it might still fall short of *guaranteeing* the truth of a belief. Think again about the bag of candy. You have pulled twenty Heath bars from the bag and one piece of candy remains. I said that you are justified in believing that the final piece of candy is another Heath bar. After all, the chance is small of pulling twenty Heath bars in a row from the bag if some other kind of candy is also in the bag. The evidence you have collected increases the odds that the last remaining piece of candy is also a Heath bar. But it doesn't have to be. By chance, it might turn out that you don't find that lonely Reese's peanut butter cup until the final draw. It comes as a delicious surprise but does nothing to change the fact that you were justified in believing that nothing but Heath bars were left in the bag. The belief you formed, that your daughter had eaten all the good candy, leaving only Heath bars, was happily false. But it was nevertheless justified.

Lotteries provide great examples of situations where people are justified in believing something that might be false. If you have bought a ticket with only a one in ten million chance of winning, you should be quite confident that you will lose. Your belief that you will lose has tremendous justification. Nevertheless, you don't *know* that you'll lose. If you knew that, why buy the ticket in the first place? Someone will win the lottery. It might be you. When it is you, you should celebrate the fact that your belief that you would lose was false. You didn't buy a losing ticket after all. But that doesn't change the fact that you had a justified belief that you would lose.

While we're on the topic, one way to see the history of science is as a history of one justified but false belief replacing another justified but false belief. Many people think that science is in the business of *proving* hypotheses, but this view is too simple. Scientists instead seek to justify their beliefs (hypotheses) about the world, recognizing all the time that their beliefs might turn out to be false. There is no absolute proof in science, which is why the oft-voiced complaint that Darwin's theory is "only" a theory cuts no ice against it. Consider, for instance, Aristotle's theory of motion. He thought that an object in motion would stay in motion only if a force continued to act on it. And, given his evidence, Aristotle's belief was justified. Newton, of course, toppled Aristotle's theory of motion, describing experiments that justified a quite different belief—namely, that forces are necessary only to *change* the motion of an object, not to perpetuate the motion of an already moving object. Newton collected a tremendous amount of compelling evidence for his theory, convincing the scientific community that Aristotle was wrong. But then, of course, along came Einstein, offering justification for yet another conception of motion.

The reason for taking all these different beliefs about motion to be justified is that all of them were backed by evidence that increased the probability of their truth. This just goes to show that the same evidence might support different theories or that it is not always possible to consider all the evidence that is necessary to show that one theory does a better job explaining a phenomenon than another theory. That some of the theories of motion were false—and perhaps one day we will find that Einstein's theory of motion is also false—doesn't mean that they were unjustified. They were not like my mother-in-law's belief about the sex of her next child or the cancer victim's belief about the healing powers of crystals.

We can now finish off the table we began earlier.

I have already explained why the belief that you bought a losing lottery ticket is false but nevertheless justified on that one occasion when you get extremely lucky and actually win. Given the incredibly tiny odds of winning, you are justified in believing that you'll lose, even if you don't lose. But the lower-right cell of table 1.2 illustrates a belief that is false and *unjustified*: that crystals have power to prevent baldness. If you own a crystal and never lose a hair, you can't credit the crystal for keeping your mop on top because no evidence exists to support the idea that possession of a crystal helps you to keep your hair. Thus, if you do buy a crystal to prevent yourself from going bald but then go bald anyway, you will know why: your belief in the crystal's power is both false *and* unjustified.

The examination of miracles in this book focuses exclusively on whether we are *justified* in believing in their existence. In terms of table 1.2, my interest is just in whether belief in miracles goes in the top row or

TABLE I.2 Truth and Justification

	TRUE BELIEFS	FALSE BELIEFS
JUSTIFIED	The next piece of candy is a Heath bar.	Your lottery ticket will lose.
UNJUSTIFIED	The next child will be a boy.	Crystals prevent you from going bald.

the bottom row. I'm not going to try to resolve the question of which *column* beliefs about miracles should go into. Unofficially, I of course have an opinion about whether beliefs about miracles are true or false, but officially, for my purposes in this book, I take no stand. We can forget about the True/False columns and think only about the Justified/Unjustified rows. The argument I make over the next few chapters is that belief in miracles belongs in the unjustified row.

I want to emphasize one last point about the significance of this difference between beliefs that are justified and beliefs that are not. I can imagine someone wondering what the big deal is. Isn't it just a matter of opinion whether a belief is justified? And if so, who really cares whether beliefs are justified? Maybe, this thinking continues, some people consider themselves to be justified in believing in miracles, whereas others don't. But there's no "objective" fact about whether beliefs are justified, so there is no fact of the matter about whether belief in miracles is justified.

I now wish to assert with complete confidence that this attitude toward justification is *very, very, very* wrong. I use the word *very* three times here because, as a philosopher, I get extremely worked up when someone dismisses a claim with the comment, "Well, that's your opinion." No doubt, some things just are matters of opinion, and when they are, I am happy to be put in my place with a reminder that what I have said is "just" my opinion. For instance, I happen to believe that chocolate ice cream is better than vanilla and that Woody Allen's early movies are much better than his more recent ones. I'm prepared to admit that these beliefs are "just" opinions. We can disagree on each point, and that would be the end of it. If you like vanilla, nothing I say will convince you that chocolate is better, and I would be foolish to insist that you're wrong and I'm right.

But when someone says that $2 + 2 = 4$ or that the planets orbit the sun rather than the earth or that the Nazis exterminated six million Jews, he or she is not simply expressing an opinion. Of course, people can disagree about these things, as ignorant or aberrant people have, but this doesn't mean that there's no right and no wrong about these issues. Two plus two really does equal four and the sun really is in the center of our solar system. If you disagree, you are WRONG. The person who denies the Holocaust can't simply "agree to disagree." The person is WRONG. The point is that not everything is simply a matter of opinion. You don't have to be shy about correcting people when they make a claim that runs inconsistent with how the world really is. Tolerance is fine in some contexts, but not in others. To say that everyone has a right to their opinions sounds good, but make sure, when you express this sentiment, that you don't take it to mean that there are never circumstances when people are simply wrong.

With that rant out of the way, let's return to the issue of justification. To the person who says that it's just my opinion that beliefs based on horoscopes are unjustified or that it's just my opinion that pulling twenty Heath bars from the trick-or-treat bag justifies my belief that the next piece of candy will also be Heath bar, I say, "NO, it is not." I am right that the first belief is *un*justified and that the second *is* justified. I am as right about this as I am that $2 + 2 = 4$. The person who disagrees with me is mistaken.

Justification, remember, raises the probability that a given belief is true, and whether a belief is true is quite often not a matter of opinion but a fact about how the world is. This means that justified beliefs "act" differently than unjustified ones, and so it is often possible to test whether a belief is justified. The most obvious test for justification will involve prediction. Beliefs that are justified should allow you to make predictions that *on average* are better than predictions you make with unjustified beliefs. If my belief about what piece of candy I will next pull from the bag is justified, my prediction should be right more often than it would be if I had no justification for believing something about the candy. This isn't to say that it has to be true. We saw earlier that justified beliefs might be false. But it is a fact nonetheless that the evidence you have collected about the candy in the bag makes your prediction about the next piece of candy more likely to be true than if you were simply guessing on the basis of no evidence at all.

In contrast, horoscopes, if specific enough so that they actually make a precise prediction (e.g., the moon will crash into the earth today!), don't *as a matter of fact* make predictions that would be any more accurate than pure guesses. I don't wish to be dogmatic about this. If my astrologer (I don't really have one) were to make a string of very precise predictions that came out true (a bird will poop on your windshield today on your way to work; your daughter will score a 98 on her algebra test; the Packers will beat the Vikings by a score of 21–17 this Sunday; the Dow will lose 3.7 percent in trading today, and so on), then I would eventually accept that her pronouncements do *in fact* justify my belief that her next prediction would be true as well. I would come to believe in astrology. But, of course, horoscopes never are so precise that they can actually be tested in a way that would justify someone's belief in them. I have no doubt that astrologers would quickly go out of business if they made their horoscopes precise enough to be tested.

Therefore, as I present my arguments against justified beliefs in miracles, I take myself to be asserting a factual claim. No one is presently justified in believing that miracles have occurred in the past.

Back to the Story about the Story about the Frog

Earlier I said that the man in the bar, Jim, believes that a frog with magical powers lives in India. I claimed that his *believing* this should be distinguished from his being *justified* in believing this. There are beliefs, and then there are justified beliefs. I haven't yet said anything about whether Jim's belief is justified or not. I have been concerned only with explaining the difference between the two. I also said that Jim's belief may be unjustified *even if* we cannot rule out the possibility of miracle-performing frogs. The tables built earlier make this idea clear. Maybe Jim's belief is true; maybe not. We don't need to make that determination to decide which *row* Jim's belief belongs in—the justified row or the unjustified row. Finally, I explained why it's not simply a matter of opinion whether Jim's belief is justified. It is either justified or not, regardless of how Jim sees things.

Now to the relevant question: Is Jim's belief in the frog justified? My answer: it is not. I hope this response does not come as a surprise. When I spoke earlier about justification, I distinguished it from mere *explanation*. I said that people might be able to explain why they believe what they do even if they can't justify what they believe. Your belief that you shouldn't leave your house today, if based on an astrologer's prediction, is unjustified. However, you might still appeal to the astrologer's warning to *explain* why you believe that you shouldn't leave the house. But what explains why Jim believes in miraculous Karnatakan frogs, and why doesn't his explanation count as a justification?

Jim admits that he has never seen the frog. He is not an eyewitness to the frog's healings and resurrections. Everything Jim knows about the frog's deeds comes by way of testimony from others, who in turn are repeating stories about the frog that they have heard from yet others. This fact—that Jim's belief is based on testimony from others   is certainly not by itself a good reason to deny that it is justified. I am prepared to admit that many of my justified beliefs are the product of what others have told me or of things I have read or of news programs I have heard on the radio or seen on TV. I believe that Antarctica exists, and I am justified in believing this although I have never seen Antarctica myself. All I know about Antarctica comes from what others in one way or another have told me about it. Likewise, although I have never met my best friend's sister, I believe that he has a sister, and I believe this only because he has told me that he does. Nevertheless, I think I'm justified in believing this.

So it is not simply because Jim's belief in the miracle-working frog depends purely on testimony that it is suspect—that it is unjustified. Why, then, is it? When does testimony fail to justify? An obvious source of failure is the unreliability of the testimony. If my best friend were a compulsive liar or delusional or a prankster, then my belief that he has a sister based on what he tells me is unjustified. Similarly, if my friend has a special interest in getting me to believe that he has a sister (maybe he wants to set me up on a date with a friend and thinks I will like her more if I believe she is his sister), then his testimony is no good. I simply can't trust what he says. Of course, I can, as we have seen, *explain* why I end up believing he has a sister—he has told me he has one. But this explanation no more justifies my belief than an astrologer's prediction justifies your belief that you shouldn't leave the house today.

But a testimony's unreliability is only one reason to dismiss its value as justification for belief. As we will see when we finally get around to evaluating the justification for beliefs in miracles, just as important is the nature of the event that the testimony reports to have happened. Even testimony that comes from a usually reliable source—testimony that under ordinary circumstances justifies you in believing that some particular event occurred—might not be enough to justify your belief in an event that, for whatever reason, is highly improbable. It's one thing for testimony to justify your belief that your best friend has a sister and quite another for testimony to justify your belief that your best friend can raise the dead, *even if the testimony is equally reliable in both cases*. But I'm getting ahead of myself. Before moving on to the arguments against justified belief in miracles, we first have to come to an understanding of what miracles are.

2

MIRACLES

WE SHOULD NOW BE FAMILIAR WITH THE DIFFERENCE between justified and unjustified beliefs. My aim so far has been the modest one of simply explaining the difference. Everyone—both believers and nonbelievers in miracles—should presumably be prepared to accept that some beliefs are justified and others not. This fact, whether we are conscious of it, is central to all of our lives. We rely on the difference between justified and unjustified belief when we choose to accept a doctor's advice rather than a shaman's or when we refuse to eat foods to which we have previously had an allergic reaction rather than trying them again on the chance that things will go better this time.

Another way to understand the significance of the justified/unjustified distinction is to reflect on the difference between beliefs that you think are true on the basis of some sort of evidence versus those that you simply *want* to be true or *wish* were true. Beliefs of the latter sort have nothing to recommend them except, perhaps, that it might be really nice if they were true. Sadly, beliefs don't become true just because you would like them to be. If they did, I would get myself to believe that a big bowl of ice cream will appear in front of me in the next ten seconds because I have a hankering for some mint chocolate chip. No doubt living in a world where beliefs were true just because you wanted them to be would have its advantages. Just imagine. You would never become ill unless you wanted to. You would become rich if you believed that you would be. You would never die, or if you did, you would live in paradise for eternity if that's what you believe would happen. If only wishing that something were true actually made it true.

To demand justification for one's beliefs is to understand that their truth doesn't rest on their desirability. Justification requires evidence of some kind. A justified belief is one that is more likely to be true than it would be without the justification. This is why, for instance, horoscopes do not justify belief—the belief you form on the basis of your horoscope is no more likely to be true than the belief you form if you ignored the horoscope. Many people believe in miracles, but are they justified in doing so? Do they have a justifying reason to believe, or do they believe just because they very much wish that miracles have happened?

My job in this chapter is to examine closely the nature of miracles. We will be trying to answer the question "What is a miracle?" It may come as no surprise to learn that over the centuries philosophers have defended different conceptions of the miraculous. Indeed, some—for example, the great Dutch philosopher Baruch Spinoza (1632–1677)—have denied that the idea of miracles even makes sense.[1] The next chapter explores why they have said this. However, so that you don't die of suspense, I can tell you now how I think we should define miracles. Miracles, I argue, should be understood as events that are the result of supernatural, typically divine, forces. Moreover, because, for reasons I soon explain in more detail, the best evidence for the presence of supernatural activity is that activity's vast improbability, we should expect that miracles are very rare. A simple idea lies behind this second claim. Supernatural events should be as far from the normal course of things as one can imagine. Thus, the more unlikely the occurrence, the more reason to believe that something supernatural is taking place.

Jim claims that the Karnatakan frog speaks several languages. It is also capable of curing the sick, causing the limbs of amputees to regenerate, and raising dead pets. Or at least dead pets that aren't ferrets. Supposing the frog to be really capable of these feats, do we wish to call them miracles? Before answering this question, let's consider some other frogs and some other things frogs do.

Frogs start their lives as tadpoles, sit on lily pads, lay eggs, swim in ponds, and, when on land, hop. Some of these activities are not too exciting, but others are pretty amazing. Consider the fact that frogs hatch from eggs, grow into tadpoles, and then end up as, well, frogs. Imagine that you knew nothing about the frog's life cycle and a friend showed you a young tadpole in an aquarium. "Believe it or not," your friend says, "this will soon sprout legs, lose its tail, and become a frog." You might not believe your friend. Surely the process your friend describes is like nothing you have seen before. Cats start as kittens, but kittens are not so different from cats. In fact, cats basically are kittens with a bit "more of the same" thrown in. Same with the relationship between dogs and puppies as well as between cows and calves. All your experience tells you that the squirmy little fish you're observing in your friend's aquarium cannot possible grow into a frog. And yet, sure enough, as the days go by, the tadpole becomes a frog.

Question: Should you think that this metamorphosis is miraculous? I contend that you should not. Let's now figure out why.

One reason for not counting the change from tadpole to frog as miraculous is that, after all, this change is *not* so different from other things you have observed. Chickens start as eggs but end up as birds. That's pretty weird when you think about it, but weird is not the same as miraculous. Similarly, butterflies were once caterpillars. Who'd have thought? So it turns out that what's happening in the case of the normal frog is not so exceptional after all. This, then, is one reason to discount the possibility that the frog's transition from tadpole to green amphibian is miraculous: it's not different from other things we have observed.

Another point worth making is that what's true of the frog in the aquarium is true of *all* frogs. Suppose you have never seen chicken eggs or caterpillars. Perhaps all your experiences with infantile forms of animals and adult forms have been like your experiences with kittens and cats, puppies and dogs, calves and cows, human infants and human adults. You might in this case be completely unprepared for what happens to the tadpole. But surely the fact that the event taking place in your friend's aquarium also takes place many, many times in ponds all over the world should dampen your enthusiasm for the idea that you have observed a miracle. What's happening to the frog in the aquarium is simply not improbable enough. It happens all the time and everywhere frogs happen to live.

In fact, your suspicion that there's nothing miraculous about the transition from tadpole to frog might spur you to investigate the natural causes that explain how it happens. If you study genetics and biochemistry and molecular biology and all the other sciences that contribute to an understanding of frog development, then you'll end up with a pretty good grasp of the sequence of perfectly natural events that suffice to turn a tadpole into a frog. Because there are no unexplained gaps in the stages of frog growth or no gaps that you have any reason to think a bit more scientific investigation couldn't fill, you can pronounce with confidence that the change from tadpole to frog is not miraculous. In fact, you now have a *justified* belief that natural causes are sufficient to explain the change from tadpole into frog because you're able to cite the laws and facts that detail exactly why tadpoles turn into frogs, and on the basis of these laws and facts you can predict with great accuracy how such changes occur, the rate at which they occur, and so on.

Having worked out why we don't think that the steps between tadpole and frog are miraculous, we can now draw some lessons about which events we should regard as miraculous and which not. Miracles should be unlike anything you have ever experienced. They should be contrary to everything you know about how the world is supposed to work. As I said, someone who didn't know that tadpoles turn into frogs might be quite surprised to see it happen. But familiarity with the way other kinds of animals grow, especially animals that undergo strange metamorphoses, such as butterflies and dragonflies, should diminish the strangeness of the case with the frog. It's perhaps incredible, but not *that* incredible.

I can summarize this idea by saying that miracles should be *extremely* improbable. The "extremely" here must be given a great deal of weight. Many improbable events occur that are not miracles. For instance, improbable coincidences occur, but their causes are mundane. The improbability of a miracle ought to be so great that it naturally leads you to think about the other aspect of miracles that I will soon be discussing—that a supernatural and usually divine hand is involved in their production.

Some care is needed to understand exactly how improbability relates to miracles. Being improbable, even extraordinarily improbable, does not *make* something into a miracle. Rather, the improbability of an event provides evidence that it is a miracle. I will explain why in a minute, but for now it's important that we distinguish evidence for a miracle from the miracle itself. The distinction is straightforward in other contexts. Suppose you wake up one morning unsure where you are. Perhaps you fell asleep on a train and have no idea how long you have been asleep or where the train was heading. You have a hunch, however, that you're in Portland, Oregon. After days of watching the rain fall, you conclude that you are indeed in Portland because it almost always rains in Portland. But if the sun has blazed brightly in the sky day after day, you are forced to conclude the opposite—that wherever you are, it ain't Portland.

We can also explain, if we wanted to, *why* heavy rainfall counts as evidence that you're in Portland. Although I know essentially nothing about meteorology, I imagine the explanation would involve facts about Portland's location in a valley near the ocean. Maybe the mountains nearby also have something to do with the amount of moisture that accumulates around Portland. For all these reasons, we expect that Portland will experience a lot of rain, and that's why seeing a lot of rain is evidence that you're in Portland.

Just as we can explain why rain counts as evidence that you're in Portland, we can also explain why extraordinary improbability counts as evidence for miracles, and this brings us to the second lesson I want to draw from the story about Jim's favorite frog. Improbability counts as evidence for miracles because miracles—at least the sort of interest to me—involve a supernatural intervention into the natural order of things. Governing nature are various laws or regularities or generalizations that, we think, must hold true. Suspension of any of these regularities requires a power outside of nature—a supernatural force. But—and now we come to why improbability counts as evidence of a miracle—we expect that violations of those generalizations that define a "natural order" must be extremely rare. They cannot be the kinds of events that happen all the time or even just now and then. They must be singular.

But why, you might be wondering, can't miracles be common? Why can't a supernatural force be constantly violating the regularities that describe natural occurrences? I agree that such constant violation is a possibility. Indeed, I have no way of proving otherwise. Perhaps every time the tide goes out, God intervenes to bring it back in. Maybe each time I successfully start my car, God provides the spark because in fact the battery under the hood has been dead for seven years. Maybe coffee is actually poisonous, and every time someone drinks a cup, God steps in to prevent what would otherwise have been an inevitable death. Although any of these things is possible, I don't know how we can accept such a possibility while at the same time making sense of the idea that some events are natural and some are supernatural. If everything that occurs is the product of God's intervention, then how should we understand the idea that there's a natural order *into* which God is intervening? Is it a natural order that never in fact unfolds? That view seems incoherent. And if miracles are the normal course of things, why regard them as *super*natural? There turns out to be no "natural" for the supernatural to be super to!

Here's another way to appreciate why we should be reluctant to regard common events as miraculous. Let's suppose that you have grown up in Indonesia, on a small island that straddles the equator. You have never seen ice. Then, for whatever reason, you decide to move to Madison, Wisconsin. You arrive in summer and notice that water in Wisconsin behaves pretty much like water in Indonesia. It pours from bottles, swallows up denser objects and allows less dense objects to float, saturates clothing, and so on. A few months later, something very peculiar occurs. You step outside of your warm house into the frigid morning and notice that the puddle of water that only yesterday appeared completely unremarkable now differs dramatically. Its surface is hard. Objects that once would have sunk to the bottom of the puddle remain on top. You can even walk across it without getting wet! It's a miracle!

Excited, you rush to your neighbor's house to tell her the good news. God has made himself manifest in the form of hard water. But your neighbor has something disappointing to tell you. She informs you of an extremely well-confirmed regularity: water freezes when the temperature drops to 32°F. Although you might at first protest (Why on earth, you wonder, should a simple change in temperature turn a liquid into a solid? Can anything be stranger than that?), through her patient clarifications you finally concede. The reason you give up the idea that the transformation of liquid water into ice constitutes a miracle is that you now accept that it is a natural occurrence. And the reason you accept it as a *natural* occurrence is that it happens all the time given the proper conditions. In fact, "it happens all the time" *is* exactly why we regard occurrences as natural. Why do we think that it's perfectly natural that a stone falls when dropped or that metal expands when heated or that days are shorter in the winter than the summer? We do so because these events and others like them happen all the time.

If a doctor heals a leper simply by laying his hands on the sufferer, we might think a miracle has taken place. But if he does this again and again, and if he opens a clinic where he teaches other doctors how to lay their hands on lepers in a way that heals them, and soon thousands of doctors are healing lepers with their hands, we would start to doubt that a miracle was involved. What once was improbable now seems commonplace, and alongside the loss of mystery comes the suspicion that the first doctor stumbled onto some kind of cure that we don't yet understand. No doubt we would begin to study how the technique he has taught the other doctors works to cure victims of leprosy. What we wouldn't do is say, "It's a miracle that happens again and again all over the world wherever these specially trained doctors happen to be."

To sum up, although I cannot prove that miracles do not happen all the time and everywhere, I think such a position threatens the integrity of a distinction that nearly everyone accepts—a distinction between the natural and the supernatural, between how things operate were God not to interfere and how things change when God decides, for whatever reason, that the ways of nature require some tinkering. Of course, those who deny the existence of God also reject the idea that God ever tinkers with natural laws. For atheists, whatever happens, no matter how strange (liquids turning into solids!), is consistent with nature's ways. A theist response—if everything were natural, we could not possibly explain, say, the dead returning to life or seas parting or staffs turning into snakes—rests on the belief that nature is not enough, that sometimes, rarely, supernatural forces must enter the scene. The believer looks to the supernatural when the natural seems inadequate. And because the inadequacy of a natural explanation is so extremely rare, the tremendous improbability of an event counts as evidence of the occurrence of something supernatural.

Let's now see how these points about miracles apply to Jim's claim that the Karnatakan frog performs miracles. First, have we ever seen a frog that speaks several languages, cures the sick, and resurrects pets? Speaking for myself, I can honestly say that I haven't. What this frog does is completely improbable. It's unlike anything we have ever seen before and seems contrary to everything we know about the abilities of frogs and the tendencies of limbs not to regenerate and dead pets not to revive. Moreover, we have no reason to believe that India is hopping with frogs like this (pun intended). If we did, then we would begin to suspect not that miracles were taking place, but that some wonderful new species of animal had been discovered. Or perhaps we would conjecture that India has been invaded by extraterrestrial beings that, despite looking like frogs, possess powers hitherto unseen by us. So the frog's behavior seems to provide good evidence that supernatural powers are at work. It does things unlike anything we have seen before. What it does is extremely improbable.

Next we must ask whether science might explain the Karnatakan frog's abilities. Although having a natural explanation of the events going on with the frog is a possibility, I confess that its likelihood seems remote. We know enough about frogs to know that they are incapable of speaking, let alone in several different languages. And although toads might give you warts, frogs certainly are not able to cause the limbs of amputees to regenerate. And resurrecting dead pets, except for ferrets? Not even the most skilled doctors can bring back the dead. If science tells us anything, it's that the dead tend to stay that way. That a frog can reverse this rule confirms the idea of something *supernatural* at work. The frog appears to have godlike powers.

Let's summarize these ruminations. I have proposed two criteria the Karnatakan frog satisfies that together make plausible the claim that if it's really doing what Jim says it is doing, it's performing miracles. The two criteria are:

1. *Extremely improbable*: a miracle should be unlike anything we have seen before. It should be contrary to everything we know about how the world works.
2. *Supernatural*: a miracle can't have a natural explanation. It must be the product of supernatural and typically divine agency.

I like these criteria. They seem to capture most examples of events that most people regard as miraculous. I also like that the two criteria work together. The first criterion, as noted, makes a point about evidence. The improbability of an event—and it had better be *very* improbable—sends us searching for a "nonnatural" explanation, a business-*not*-as-usual sort of thing. The second criterion describes the feature that actually makes an event miraculous—its supernatural origin.

You have probably noticed that I tend to modify my description of a supernatural cause with the phrase *typically divine*. I have done this because many people who believe in the supernatural seem not to take the actions of any old supernatural agent to be miraculous. When a ghost roams their attic late at night rattling chains, most people, I think, would deny that this activity constitutes a miracle. The cacophonous racket has a supernatural cause, and it may very well be extremely improbable (for sake of a good night's sleep, I hope it is!), but because the cause isn't divine, many would be reluctant to describe the event as miraculous.

Likewise, possession by the devil clearly involves a supernatural cause, but no one wants to say that when the devil possessed poor Regan in the movie *The Exorcist* (1973), a miracle had been performed. Most of us intuitively conclude that miracles come from God, and that's why we usually conceive miracles as signs from God. However, I don't want to insist on this point. If you think that miracles might have supernatural but nondivine origins, that's OK with me. My argument against justified belief in miracles requires only that miracles have supernatural causes of some sort, although I confess that the argument goes through more directly when these supernatural causes are thought to involve divine agency.

In closing this chapter, I would urge that no one should dispute that many of the (alleged) miracles that have most impressed people over the millennia satisfy my two criteria. If Moses parted the Red Sea, this was an extremely improbable event, making us favor the idea that it must have had a supernatural cause. Likewise if Aaron turned his staff into a snake or if Jesus raised Lazarus from the dead. So even if you think that my criteria fail to capture everything that should count as a miracle or perhaps encompass events that should not count as a miracle, I'm not too concerned. I'm after the big fish, and they happen to be events that are so extremely improbable that we seem compelled to attribute supernatural, typically divine, causes.

It's now time to get to business. If I'm right that miracles are vastly improbable because of their supernatural origins, then an argument against being justified in believing in miracles has two openings. The argument might focus on the improbability of miracles, showing why one cannot be justified in believing in the occurrence of events as improbable as miracles. Alternatively, the argument might concentrate on the inference from miracles' improbability to their supernatural origin, finding fault with claims to have justified beliefs about supernatural causes. In the chapters that follow, I develop *both* kinds of argument. In chapter 3, I explain why we cannot be justified in believing that some event has a supernatural, typically divine, cause—even if the event is as improbable as miracles purportedly are. I think this argument is fairly straightforward but completely convincing. The second argument, centering on the improbability of miracles, involves more complications and must be made in two stages, so it will be elaborated over three chapters. By the end of chapter 6, then, I will have made the case that no one presently has or has ever had justified belief in miracles.

3

JUSTIFYING BELIEF IN

SUPERNATURAL CAUSES

IN THIS CHAPTER, I CONSTRUCT MY FIRST BIG ARGUMENT against the idea that we can have justified beliefs in miracles. The problem I focus on concerns the idea that miracles should arise as a result of supernatural and typically divine forces. I come to the conclusion that there is no way to justify beliefs about the supernatural origins of those events that are regarded as miracles, so there is no reason to be confident that what we are witnessing when we *think* we're seeing a miracle really is a miracle.

As I mentioned, this chapter contains the *first* big argument against justified belief in miracles. There's a second big argument in the wings, but that's the stuff of the next three chapters. However, like the first, it follows from the conception of miracles that I developed in chapter 2. Whereas the argument against justified belief in this chapter focuses on the part of the definition of miracles that involves the supernatural, the argument in the next three chapters concentrates on the idea that miracles are extremely improbable.

Identifying Causes

The feature of miracles that is crucial for the rest of this chapter is their supernatural origin. If you don't think that miracles must have supernatural causes, then the argument that follows will not be of any concern to you, and you're free to continue to think that your belief in miracles might be justified (that is, until you read the next three chapters!). However, I do wonder about the idea of miracles having natural causes. Why call an event a miracle if it has a natural explanation? This is why eclipses, rainbows, and earthquakes, as improbable as they seem to be (surely they're more probable than miracles, anyway), don't count as miracles.

Trying to say precisely what a supernatural event is turns out to be rather difficult. I think most people, whether through exposure to childhood fairy tales involving witches or to movies in which ghosts play prominent roles or to novels about wizards or vampires, have no trouble imagining the possibility of supernatural occurrences. These occurrences count as *supernatural* only if they have no explanation in terms of the natural world. When a witch straddles a broom and flies through the sky, she's doing something that can't really happen if only natural laws apply to the world. From what we have learned in Physics 101, flying on a broom shouldn't be possible. Gravity should prevent the broom from hanging in the air for more than an instant unless it has some sort of natural external propulsion.

Likewise, we know enough about the physiology of animal bodies and the capacities of frogs to be quite confident that Jim's Karnatakan frog is doing things that would be impossible unless something supernatural were occurring. Natural just ain't enough to regenerate dog and cat limbs or to bring dead pets back to life. This is one reason, I think, for accepting that the frog, if it really is doing everything Jim reports, is performing miracles. If nothing in nature can explain what the frog does, its powers must be more than just natural. They must be supernatural.

All this sounds right, but when you start thinking philosophically about what it means for something to be supernatural or how you might know when something has a supernatural cause, matters get messy. Good reasons exist for resisting the idea that supernatural events involve *violations* of the laws of nature. If a law is truly a law, it *cannot* be violated. If it is violated, it wasn't a law in the first place, and so the "miraculous" event involves no violation of a law. I discuss this reasoning further in appendix 1. Questions also arise about how something that's genuinely supernatural can cause things to happen in the natural world. Because discussion of this problem would take us too far afield, I have included it in appendix 2 for the interested reader.

However, let's pretend these problems don't exist. We will assume that if an event is supernatural, it must have its origins in something outside nature, and whatever this something is, it must be causing something else to happen that wouldn't be happening if only natural laws were in operation. Thus, when Jesus walked on water or turned water into wine or raised the dead, he was obviously performing actions that couldn't occur if the world were working the way that it is "supposed" to be working—the way it works when only the laws of nature are in effect. Similarly, when statues of Mary weep, or when stigmata appear on the hands of believers, or when the Red Sea parts, we're right to think that the normal, natural course of events that would ordinarily have made these kinds of things impossible must have been interfered with in some way. Something *unnatural*, or *supernatural*, is going on.

As should be clear from these few words about supernatural forces, we come to believe in them as a result of a specific kind of *inference*. The inference goes like this: You observe something the likes of which you have never seen before. It's something that should be impossible given everything you know about how the world works. You then conclude that the cause of what you're observing cannot be natural. It must be supernatural.

It should be fairly obvious why we need to *infer* the existence of the supernatural, but the point is worth making explicit. The problem with supernatural causes is that you can't directly observe them. When you see one billiard ball knocking into another and causing it to move, you don't need to *infer* that the first ball caused the second to move. You see that directly. You *witness* the first ball hitting the second. The cause of the second ball's motion is plain as day.

But when you see a statue of Mary crying, you can't see what causes the statue to cry. You have to make an inference. You reason like this: From what you know about the natural world, marble statues can't cry. Therefore, the cause of the statue's crying must be something supernatural. Likewise, from what you know about bodies that are denser than water, they shouldn't float. So when you see someone walking on water, you reason in the same way you do regarding the crying statue. You don't see what causes the person to walk on water, and you know that nothing natural can explain what you see, so you infer that a supernatural cause must be doing the work.

There's nothing really suspicious about an inference of this sort. We use it not only when identifying supernatural causes but also when identifying anything that we can't, for whatever reason, directly observe with our senses. For instance, when I head downstairs tomorrow morning, I will find a newspaper on my stoop. I don't actually see how it gets there. The newspaper arrives before I do, at least when everything is going the way it's supposed to go. So what caused it to be on my stoop? Even though I don't observe the newspaper deliverer, I do see the newspaper and infer from its presence that the deliverer has been by. Likewise, I don't see the gasoline in the tank of my car, but I'm on safe ground in inferring that the gasoline is present because the car starts when I turn the key in the ignition. The only difference between inferences like these and inferences that lead you to claim the existence of supernatural causes is an extra step—the step that says "and nothing natural can be causing this event." It is the "and nothing natural . . . " premise that forces you to the conclusion that supernatural forces are at work. You simply don't need that premise when wondering how the newspaper found its way to your stoop or how the engine in your car turned over. Although you haven't directly observed the causes in these latter cases, you needn't assume that the supernatural caused what you're presently observing.

Inference to the Best Explanation

I want to spend more time examining the pattern of inference that we use whenever we are trying to identify causes that we can't directly observe. Having a clear understanding of the limitations of this form of inference is crucial to my first argument against justified belief in miracles. Because philosophers argue with each other for a living, they have spent a great deal of time categorizing and studying inferences and arguments (I use the words *inference* and *argument* interchangeably here). They do this so that they have a better understanding of how and the extent to which reasons support a conclusion. All this thinking about arguments presumably qualifies philosophers to judge when an argument is weak or when one argument is stronger than another. It's not always an exact science, but you might be surprised at the progress philosophers have made in understanding the various forms that inferences might take.

The kind of inference that allows you to identify causes that you can't directly see for yourself is what philosophers call an *inference to the best explanation*. Another name for this kind of argument that you might see is *abduction*. I like the first label better because it provides a concise description of the kind of reasoning that the argument depends on and doesn't make you think about kidnappings. Let's first look at an example of an inference to the best explanation, and then I'll make some important clarifications.

The setting: A kitchen with tile countertops. Sitting on the countertop, beneath some white kitchen cabinets, is a cookie jar. The jar's lid sits on the counter next to it. Spread about on the counter are cookie crumbs. Chocolaty fingerprints blemish the otherwise clean surfaces of the kitchen cabinets. The cookie jar is empty.

The description of the setting I have just provided comprises what I shall call *observations*. Observations provide the evidence or data that we'll use to help us in our pursuit of the cookie thief. We have two suspects. Nancy, the older sister, is usually pretty tidy and takes care to clean up after herself. Miranda, the younger sister, always needs reminding to put away her things and to wash her hands after eating. I now wish to consider two hypotheses about who stole the cookies. The first hypothesis, which I'll call *Neat Nancy*, is that Nancy stole the cookies. The second hypothesis, *Messy Miranda*, is that Miranda stole the cookies.

We now want to put our observations to work in helping us decide which of the two hypotheses to believe. The way we do this is to think about the connection between the hypotheses and the observations. One of the hypotheses, *Neat Nancy*, makes the observations surprising. From what I have told you about Nancy, we can conclude she would have replaced the lid of the cookie jar, would have cleaned up any crumbs she left on the counter, and would have washed the chocolate from her hands before opening the cabinet. In contrast, the observations fit just what you would expect if the *Messy Miranda* hypothesis were true. Because Miranda tends to be lazy and sloppy, it comes as no surprise at all that we would see crumbs and fingerprints if she were the cookie thief. The *Messy Miranda* hypothesis gives us a better *explanation* of the observations than the *Neat Nancy* hypothesis. Inference to the best explanation leads us to prefer *Messy Miranda* to *Neat Nancy*.

In this example, I quite deliberately chose to use the jargon *observations* and *hypotheses*. This usage was meant in part to highlight the fact that inference to the best explanation is pervasive in scientific investigations. Scientific discoveries often concern things that no one can directly observe. How, then, can scientists be justified in believing that these things exist? The answer is inference to the best explanation. Thus, for instance, the British scientist who discovered electrons in 1897, J. J. Thomson, never actually saw an electron. They're too small to be observed. Rather, he relied on inference to the best explanation to support his hypothesis that electrons exist. He drew his observations from a series of experiments in which he passed a current through a glass tube and observed the behavior of the resulting glowing rays of light (cathode rays). On the basis of these and some other observations, Thomson made an inference to the best explanation—that the rays consisted of charged particles that we now refer to as electrons.

Similarly, in 1847 Ignaz Semmelweis could not directly observe the bacteria that obstetricians carried on their unwashed hands when they delivered babies in a hospital in Vienna. But when he insisted that they wash their hands after dissecting cadavers and before entering the obstetric ward to deliver babies, he saw the mothers' mortality rate drop from around 20 percent to around 2 percent. This put him in a position to infer that the doctors had been transferring invisible germs to the mothers.

Sometimes the label *inference to the best explanation* forces one to accept a grammatical infelicity. The word *best* implies that more than two explanations compete with each other to make sense of a set of observations. However, we might have only two competing explanations, or hypotheses, in mind, in which case what we're trying to do is to infer to the *better* explanation. Thomson's hypothesis about electrons, for instance, was in competition with another hypothesis that took the luminescence in the glass tube to be caused by the actions of waves rather than particles. Likewise, Semmelweis's hypothesis that invisible germs carried disease competed with the idea that disease arose spontaneously in some people. The point is that for inference to the best (or better) explanation to yield results, you need to consider more than one hypothesis to explain the observations.

Sometimes the competing hypotheses are so vague that you can't even state them explicitly. For instance, I earlier mentioned inferences to supernatural causes. The statue of Mary weeps. What causes this? One hypothesis is that a supernatural cause is at work. The competing hypothesis is that the cause of the statue's weeping is something natural, but you can't say what that something is. We will see soon the special troubles that come along with inferences to supernatural explanations.

Here's another point to make about inferences to the best explanation. An explanation that does a better job explaining some set of observations when compared to another explanation or that does the *best* job when compared to two or more others may eventually lose its claim to be the better or best explanation. As more observations become available, a formerly "best" explanation might end up not as good as some new explanation. Suppose, on closer inspection of the scene in the kitchen, you notice muddy footprints on the floor that match the tread on the sneakers of Miranda's sloppy friend Margaret. The footprints start at the front door, make a beeline for the cookie jar, and then fade from view as they move toward the back door. You now have a third hypothesis to consider: *Meandering Margaret*. The latter seems to be the best hypothesis, better than *Neat Nancy* and *Messy Miranda*, because it explains more observations than either of these other two hypotheses. Before you notice the footprints, inference to the best explanation led you to favor *Messy Miranda* over *Neat Nancy*. Now, with additional observations, you like *Meandering Margaret* best of all.

I now want to say something about inference to the best explanation that will figure very significantly in our future discussions of miracles. Sometimes an inference to the best explanation is not possible. This will happen when the hypotheses under examination do an equally good job of explaining the observations. For instance, suppose no footprints muddied the kitchen floor. *Neat Nancy* and *Messy Miranda* are the only two hypotheses under consideration. But let's now introduce some new facts about Nancy. Maybe Nancy is cunning and jealous of all the love her parents lavish on her little sister. She decides that she will frame Miranda for the unpardonable crime of cookie theft. Knowing Miranda's habits, she arranges the kitchen to look exactly as it would if Miranda were to steal the cookies. She leaves the cookie jar lid on the counter, scatters crumbs about, and smudges the cabinets with fingers.

When you return home from work, you look with horror at the empty cookie jar and the accompanying mess. But which hypothesis should you prefer? Should you infer that sloppy Miranda stole the cookies or cunning Nancy? Neither hypothesis makes the observations surprising. *Both* lead you to expect the very observations that now lie before you. In such a case, inference to the best (or better) explanation is of no use to you. Of course, the observations still suffice to rule out other hypotheses—for example, that your very sweet and tidy spouse stole the cookies. But relative to *Messy Miranda* and *Cunning Nancy* the observations don't license an inference to the better explanation. The thing to do is to collect further observations that would drive you to prefer one of the hypotheses over the other. Perhaps you could interrogate the suspects. If Nancy confesses, this confession would tip you toward the *Cunning Nancy* hypothesis. Likewise, a confession from Miranda would steer you toward the other hypothesis. (Or, if this were an episode of *CSI: Crime Scene Inves-*

tigation, you could take their fingerprints and compare them with the fingerprints left at the scene of the crime, which would definitively lead you to one hypothesis or the other.) But when you are faced with observations that are equally expected given the hypotheses at hand, inference to the best explanation *is of no help*.

Inference to the Best Explanation and Background Assumptions

I'm afraid I need to complicate the discussion of inference to the best explanation, but just a little bit. The problem is that inferences to the best explanation actually involve more than the comparison of two or more competing hypotheses. The real story is more complex. Think again about the empty cookie jar and the first two explanations we considered to explain it: *Neat Nancy* and *Messy Miranda*. The reason we preferred *Messy Miranda* to *Neat Nancy* was that we saw a mess around the cookie jar, and we believe that messes are usually caused by messy people and are usually not caused by neat people. This is a very obvious point, to be sure, but it's an assumption that is very important to the success of our inference. Here are some additional obvious points: we believe that cookies leave crumbs behind; we believe that chocolate melts on fingertips; we believe that chocolate leaves visible smudges on white cabinets; and so on. Why do I mention all these obvious facts?

The reason to list all of these apparently trivial facts about what messy people do and about what happens when cookies break and about how chocolate behaves when it comes into contact with warm flesh and white surfaces is that they all are in play—albeit in the background—when you make your inference to *Messy Miranda* rather than *Neat Nancy*. To see this, suppose you were uncertain of these facts. What if you didn't believe that messes are made by messy people or that cookies make crumbs or that chocolate makes smudges? Suppose you believed, for whatever reason, that storks routinely drop crumbs on kitchen counters. Or suppose you believed that the smudges on the kitchen cabinets were in fact giant brown amoebas. If you believed these things, then your observations wouldn't lead you to prefer *Messy Miranda* over *Neat Nancy*. The point is that the *Messy Miranda* hypothesis is a better explanation than the *Neat Nancy* hypothesis *only relative to your background assumptions*. Usually, of course, the background assumptions on which we rely when weighing various explanations are so obvious and so uncontroversial that we don't even realize that they're playing a role in our inferences. But they are.

Let's try to summarize the various points I have been making about inference to the best explanation with the help of a diagram.

Figure 3.1 shows how the pieces of an inference to the best explanation fit together. On the right are the observations that we're trying to explain in order to apprehend the cookie thief. Because we did not actually witness the theft taking place—because, that is, the *cause* of the theft is beyond our present powers of observation—we must rely on an inference to identify that cause (i.e., who the culprit is). We entertain two hypotheses, each of which, *together* with our background assumptions, might explain the evidence available to us. However, *Messy Miranda* clearly does a better job than *Neat Nancy* of explaining the observations, given our background assumptions, and so it is *Messy Miranda* that we

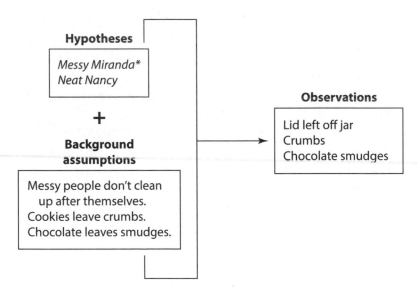

FIGURE 3.1 Messy Miranda versus Neat Nancy

are *justified* in believing. The asterisk next to *Messy Miranda* indicates that it is the winning hypothesis. Easy, right?

But now let's look at another case I described, where *Messy Miranda* is in competition with *Cunning Nancy*, who has left behind evidence to frame her notoriously sloppy sister after having absconded with the cookies herself (figure 3.2).

In this case, because both hypotheses, relative to the background assumptions, explain the observations, any preference for one over the other would be *unjustified*. You simply have no reason to choose one rather than the other. What to do? You need to collect further observations of a sort that will enable you to distinguish one hypothesis from the other. For instance, you might measure the height of the smudges on the cabinet. If among the items in your background assumptions are facts about Miranda's and Nancy's height, and if the smudges on the cabinets are above Miranda's reach, you now have a reason to prefer *Cunning Nancy* to *Messy Miranda* (figure 3.3).

In short, whenever hypotheses make the same predictions—explain the same observations—you need to hunt around for more observations

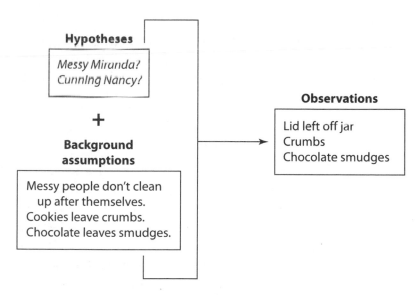

FIGURE 3.2 Messy Miranda versus Cunning Nancy

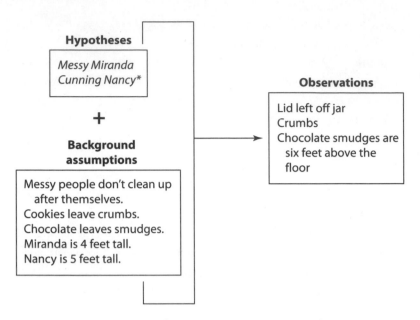

FIGURE 3.3 Messy Miranda versus Cunning Nancy, take two

and perhaps rely on more background assumptions if you are to justify your belief in one hypothesis rather than another.

Having covered the basics involved in inference to the best explanation, we're finally ready to move on to my first argument against justified belief in miracles.

Battling Serpents

More than three hundred years ago the British philosopher John Locke (1632–1704) offered a discussion of miracles in his essay *A Discourse of Miracles* (published posthumously in 1706).[1] I'm far less interested in the definition of miracles he provided in this essay than I am in his use of an inference to the best explanation to justify belief in miracles. This will be our first opportunity to see how inference to the best explanation falls short of justifying belief in miracles. And because inference to the best explanation seems to be the only, or at least the best, way to justify belief in things that we can't directly observe, this missing of the mark is a big problem for people who think they are justified in believing in miracles. After showing the specific problems with Locke's argument, I generalize these problems so that they apply to beliefs about all miracles.

Among the issues Locke discusses in his essay is a question about how to identify a miracle. Nature is of course full of amazing spectacles. Years ago I had the pleasure of seeing the northern lights when I visited friends in upstate Wisconsin. The green and purple bands that stretched from the horizon far up into the night sky were astonishing, and I will never forget them. But they weren't a miracle. How do we know a miracle when we see one? Locke's idea was that miracles have to be so spectacular, so shocking, that only a supernatural being could be their cause. Put in our terms, when an inference to the best explanation leads you to prefer a supernatural explanation of some event to an explanation that mentions only natural causes, you're looking at a miracle.

To illustrate his idea, Locke mentions the story from Exodus in which Moses and his brother Aaron confront Pharaoh to deliver word of God's desire that the Israelites be set free (Locke seems to have forgotten Aaron's part in the story and attributes to Moses the actions that Aaron performed).[2] To prove to Pharaoh that they are messengers of God, Aaron throws his staff on the ground, and it becomes a serpent. Not to be outdone, Pharaoh has sorcerers in his employ turn *their* staffs into serpents. That Pharaoh's sorcerers could perform a feat no less shocking than what Aaron did casts doubt on the omnipotence of the power working through Aaron. How do we know that God is responsible for the fantastic transformation of Aaron's staff into a snake if Egyptian sorcerers can accomplish the same thing? Perhaps Pharaoh's sorcerers are just really good magicians with no supernatural abilities at all. Perhaps they have mastery of some supernatural forces, but nothing on the level of God. Aaron's "trick" provides no evidence of anything greater. Pharaoh is left with no reason to think that Moses and Aaron speak for God.

Putting this question in terms of an inference to the best explanation helps to understand Locke's answer. Remember the difficulty that the *Cunning Nancy* hypothesis presented to the discovery of the genuine cookie thief? Given the observations of cookie crumbs and smudged cabinets, there was no reason to prefer the hypothesis that messy Miranda stole the cookies to the hypothesis that cunning Nancy stole them. Both hypotheses explained the observations. And just as we resolved that problem by seeking new evidence—evidence that one hypothesis but not the other could explain—so too Moses and Aaron must present new evidence to Pharaoh that will compel him to prefer the *God* hypothesis to the competing *Magic* hypothesis.

OK—perhaps turning a staff into a snake isn't a big deal after all. From such an observation, you don't know whether to infer the existence of God or some lesser power like the kind that Pharaoh's sorcerers possess. So then what does Aaron have to do to tip the scales in his favor—to make inference to the best explanation point toward God and away from magic? What if the serpent he created *eats* the serpents that the Egyptian sorcerers created? On the one hand, if you thought that the power behind Aaron is no greater than the power of Pharaoh's sorcerers, then you shouldn't expect Aaron's snake to be able to eat the sorcerers' snakes. On the other hand, if you thought that an *omnipotent* force is acting through Aaron, you would expect his snake to kick the butts of the sorcerers' snakes (Do snakes have butts?), and that's just what happens. Aaron's serpent eats the sorcerers' serpents, and woe unto Pharaoh for not taking the message to heart.

Now that Locke's reasoning is evident, the time has come to show where it goes wrong. Let's return for a moment to the sordid tale of cookie thievery. In order to choose between the *Messy Miranda* and *Cunning Nancy* hypotheses, we need to collect additional evidence. Absent fingerprint identification, confessions from one or the other suspect would of course be ideal, but we might look for other kinds of evidence, too, such as the height of the fingerprints on the cabinets. We might also need to draw on more of our background assumptions about cookies, chocolate, the heights of Miranda and Nancy, and so on. Moses and Aaron, as Locke points out, seem to employ a similar strategy. If simply turning a staff into a serpent isn't enough to indicate the existence of God, because such a thing could be caused to happen through other means, then additional observations, drawing on additional background assumptions, are necessary. Figure 3.4 illustrates this idea.

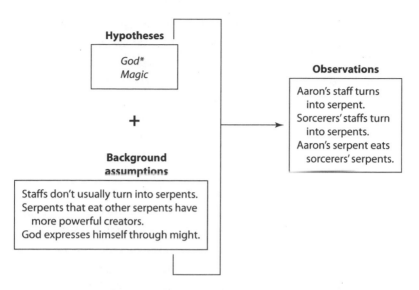

Hypotheses

God*
Magic

+

**Background
assumptions**

Staffs don't usually turn into serpents.
Serpents that eat other serpents have
    more powerful creators.
God expresses himself through might.

Observations

Aaron's staff turns
    into serpent.
Sorcerers' staffs turn
    into serpents.
Aaron's serpent eats
    sorcerers' serpents.

FIGURE 3.4 God versus magic

According to the *God* hypothesis, God is acting through Aaron. According to the *Magic* hypothesis, Aaron's action is no greater than that which Pharaoh's sorcerers performed and so gives Pharaoh no reason to think that Aaron and Moses are God's messengers. Locke thinks that an asterisk belongs next to the *God* hypothesis because he also believes the background assumption that a more powerful serpent, as Aaron's clearly is, means a more powerful creator, and God would show his power through his savage might.

Troubles for Locke

Once Locke's reasoning is made absolutely explicit, finding where it goes wrong is not hard to do. Actually, "where it goes wrong" puts matters generously because it goes wrong in more than just one place. Let's start with the background assumptions. How solid are the assumptions in figure 3.4? That is, how certain of them can we be? The first assumption seems pretty safe. I have never seen a staff turn into a serpent, and I bet you haven't either. The second assumption seems less certain. Presumably, serpents that exist in nature—what we would more commonly call snakes—don't have creators at all (except their moms and dads), and so when one snake eats another, we learn nothing about the power of their creators. And even in the case where the serpents really have been created, why should we think that a more powerful serpent entails a more powerful creator? Maybe some creators have hardly any power at all, but they make damn good serpents. Maybe I make better flourless chocolate cake than the president of the United States does. Am I thereby more powerful?

However, the least supported assumption of all in the background-assumptions box is the last. How does Locke know that God expresses himself through might? How could Pharaoh have known that? How can anyone know this? God's intentions, desires, habits, and so on are simply not available to us. Whatever we assume about God's nature is purely speculative—guesses, really. God *might* wish to reveal himself through manifestations of strength, but maybe not. Maybe God reveals himself through displays of nonviolence and pacifism by turning the other cheek when confronted by a tyrant. If this were true, then when Aaron's serpent eats the sorcerers' serpents, we have evidence *against* the God hypothesis. A pacifist God would never resort to such infantile and barbaric one-upsmanship.

So the first reason Locke's inference to the best explanation fails is that it depends on a background assumption that can't be justified. Notice that other background assumptions that we have discussed, such as that cookies leave crumbs or that Nancy is taller than Miranda, *can* be verified. This is why we don't need to think much about them when testing hypotheses about the cookie thief. Anyone who doubts whether cookies leave crumbs can make a batch of chocolate chip cookies and perform the relevant experiments; anyone uncertain whether Nancy is taller than Miranda can measure each of them. But how do we verify assumptions about God's characteristics and "personality"? When Locke assumes that God expresses himself through might, how would he go about verifying this assumption to someone who doubts it? We don't have in our possession the equivalent of a God baseball card on which God's stats are revealed (created universe in six days; becomes angry when idols are worshipped; makes very powerful serpents). This means that Locke is not justified in believing his assumption about God, which in turn means that he is not justified in believing that God was working through Aaron. But then whether Aaron actually performed a miracle can't be determined.

I mentioned that Locke's inference goes wrong in more than one place. We have so far been focused on an error that emerges from the background-assumptions box. But I now wish to consider a difficulty that stems from the hypotheses box. I have been presenting inferences to the best explanation as involving a competition between two hypotheses: *Messy Miranda* versus *Neat Nancy*; *Messy Miranda* versus *Cunning Nancy*; *God* versus *Magic*. As I noted, when two hypotheses are compared, inference to the *best* explanation might better be described as inference to the *better* explanation. But, of course, there are cases when you actually want to discover the *best* explanation—that is, when you want to compare the virtues of many hypotheses. If I wanted to, I could have listed lots of hypotheses to explain the empty cookie jar. I did suggest another in passing, *Meandering Margaret*, which involved the idea that Miranda's friend stole the cookies. But why not consider still others, such as *Curious George* or *Hungry Goldfish* or *Cookie Monster*? Of course, the reason not to entertain any of these other possible explanations for the missing cookies is that they are so obviously unlikely. Curious George is a fictitious monkey, goldfish don't like cookies and wouldn't be able to open a cookie jar, and Cookie Monster has, sadly, given up sweets for more healthful snacks. But there might in fact be *better* explanations for the empty cookie jar than the two hypotheses I did consider. Perhaps *Ravenous Rachel* does a better job explaining the evidence than either *Messy Miranda* or *Neat Nancy*. If so, then even though I am justified in believing that Miranda was the thief, I would be even more justified in believing that Rachel was the thief if only I had thought to consider the possibility. Remember a lesson from chapter 1: being justified in believing something doesn't mean that your belief must be true. It means only that you have support for your belief. My belief in *Messy Miranda* does have support. That's why it's preferable to *Neat Nancy*. But it might in the end have less support *or no greater support* than belief in *Ravenous Rachel*.

We came across this idea of *no greater support* earlier when contrasting *Messy Miranda* with *Cunning Nancy*. Until additional observations are made, *Messy Miranda* has no greater support than *Cunning Nancy*, and vice versa. The two hypotheses have equal support. When hypotheses have equal support and you have exhausted all possible observations, then inference to the best explanation is useless. It can't help you. When one hypothesis has no more support than any other, there's no such thing as an inference to the *best* explanation.

With these points in mind, let's now turn to the hypotheses box in figure 3.4, illustrating Locke's reasoning. When Aaron turned his staff into a serpent, which then ate the serpents of the sorcerers, Locke considered only two hypotheses that might account for this surprising turn of events. The *God* hypothesis credits Aaron's actions to a divine cause. The *Magic* hypothesis attributes Aaron's actions to the same kind of powers that Pharaoh's sorcerers possess—whether supernatural or not doesn't really matter in this context. Then, given the background assumptions that I have discussed, inference to the best explanation leads you to prefer the *God* hypothesis.

But now let's open our minds a bit. Why should we be satisfied with just two hypotheses? Obviously, not any hypothesis that explains Aaron's trick with the staff is worth our consideration, just as we would be wasting our time to think hard about the *Curious George* or *Cookie Monster* hypothesis when trying to discover the identity of the cookie thief. But just as we might have missed an opportunity by failing to consider *Ravenous Rachel*, we want to be sure that we truly do have in our possession the best possible slate of hypotheses for explaining how Aaron managed to turn his staff into a super-duper serpent. Here are some additional hypotheses to think about:

1. *Seventeen Gods*: there exist seventeen gods, who, working together, turned Aaron's staff into a serpent that then ate the others.
2. *Almost God*: there exists one enormously powerful but not quite omnipotent being who turned Aaron's staff into a serpent capable of eating other serpents.

3. *Supermagician*: Aaron was a very good magician, much better than Pharaoh's sorcerers, and he made it seem as if his staff turned into a serpent although it didn't really, and the serpent he used for his trick was able to eat the sorcerers' serpents.

4. *Mass Hypnosis*: Aaron was an excellent hypnotist, and he made Pharaoh believe that he saw Aaron's staff turn into a serpent and then eat the sorcerers' serpents, but it didn't really.

These four hypotheses do as good a job explaining Aaron's feat as the initial *God* hypothesis. If any of them were true, we would expect to observe exactly what Pharaoh did observe (or thought he observed): a staff turning into a serpent and then eating the serpents that the sorcerers had created. And, of course, there are plenty more hypotheses where these came from. How do we dismiss them as inferior to the *God* hypothesis? They seem silly, I grant. But they're not silly in the way that the *Curious George* and *Cookie Monster* hypotheses are silly. We *know* that Curious George and Cookie Monster don't exist. That's why we don't take seriously the possibility that either of them stole the cookies from the cookie jar. But why is *Seventeen Gods* or *Almost God* sillier than *God*? We can't say that it's because we *know* that seventeen gods don't exist and that an almost god doesn't exist. We don't know these things. Remember, belief that God is the cause of Aaron's deed is supposed to come as a *conclusion* from an inference to the best explanation—God is the explainer! The point here is that inference to the best explanation no more justifies the belief that God acted through Aaron than it justifies the belief that seventeen gods or an almost god did. There is no best explanation.

I mentioned earlier that I think miracles should be conceived as having a divine cause, but I'm not stuck on that view. Perhaps supernatural causes of any sort suffice. In that case, *Almost God* is good enough to justify the claim that Aaron had performed a miracle. But hypotheses 3 and 4, *Supermagician* and *Mass Hypnosis*, don't license the miracle claim, and they aren't silly at all. If you've ever seen the magician David Blaine at work, you would have observed him do tricks far more astounding than turning a staff into a snake. (Really, is that trick much more impressive than pulling a rabbit from a hat?) And I once attended a show by a hypnotist who was able to put half the audience into a trance and make them believe that they were on an airplane.

So Locke appears to have disregarded (or not thought of) innumerable hypotheses, each no less supported by the observations than the *God* hypothesis. For this reason, inference to the best explanation doesn't justify his belief that Aaron's actions reveal a divine presence or even a lesser kind of supernatural presence, and if, as Locke supposes, miracles are brought about only through God's efforts, then he has no justification for believing that Aaron had performed a miracle.

Of course, just as happened when we weighed *Messy Miranda* against *Cunning Nancy*, we might hope to gather further observations that will winnow the number of possible explanations for Aaron's fantastic doings. Unfortunately, time travel being what it is, we're in no position to go back to Pharaoh's court to examine more carefully Aaron's encounter with Pharaoh. Were we able to do this, perhaps we could rule out *Super Magician*. Suspecting that Aaron is nothing more than a showman, we might look around for mirrors or other tell-tale signs of trickery. Finding none, we could then reject *Supermagician*. Similarly, we might look for evidence that Aaron is a practiced hypnotist. Maybe he carries spiral disks in his pockets or a pocket watch on a long gold chain. If we find no such evidence, we could reject *Mass Hypnosis*. But what evidence could ever lead us to prefer *God* over *Seventeen Gods* or *Almost God*? What observation distinguishes an all-powerful God from a God just slightly less than all powerful or a single God working on his own from seventeen gods who can do anything if they cooperate? None that I can think of.

Generalizing the Arguments

My attention has been centered on Locke and his discussion of the battling serpents. However, showing how to apply the lessons from this discussion to belief in miracles of any sort is a pretty straightforward matter. Anytime you rely on inference to the best explanation to argue for the existence of an unobserved cause, you need to think about the plausibility of the background assumptions and the multitude of hypotheses that could make sense of the observations. Locke makes a mistake with respect to each of these considerations. In the first place, we saw, he helps himself to an assumption that he can't back up. God's intentions and desires are simply not available to us, which explains the common thought that "God works in mysterious ways." In the second place, Locke fails to take seriously alternative hypotheses that are no worse than the one he likes. *God* would certainly explain Aaron's actions, but so too would *Seventeen Gods, Almost God, Supermagician,* and *Mass Hypnosis.* Inference to the best explanation ends up unable to justify the conclusion that it is God who acts through Aaron.

But now when we take a step back and think more broadly about all reported miracles, not just Aaron's, problems parallel to those that Locke faced rise quickly to the surface. To be sure, applications of inference to the best explanation will differ somewhat depending on the miracle that you wish to contemplate. The background assumptions that allow you to infer the presence of a supernatural force given the observation of a weeping statue of Mary will differ from those you depend on to infer a supernatural cause of a dead person's resurrection. Nevertheless, we can provide a *sketch* of the kind of problem that will always present itself to inferences that try to establish the existence of a certain kind of supernatural cause, whether God, an angel, a spirit, or something else.

Each of the boxes in figure 3.5 can be filled in to accommodate a particular miracle report. We saw already how the boxes get filled with respect to Aaron's purported miracle. But let's try another on for size: the weeping statue of Mary located in Japan known as Our Lady of Akita (the only Vatican-certified instance of such an event). Into the observations box goes a description of this extremely improbable occurrence: a wooden statue of the Virgin Mary appeared to weep on a number of occasions in the 1970s. We then toss several explanations for this observation in the hypotheses box: a supernatural force causes the statue to weep; the weeping is the result of a natural cause that we don't presently understand; the statue weeps because extraterrestrials have aimed a weeping beam at it. Next, we fill the background-assumptions box: wooden statues tend not to weep; if the Virgin Mary (or some other supernatural power or an extraterrestrial) wants to reveal herself (itself) for purposes

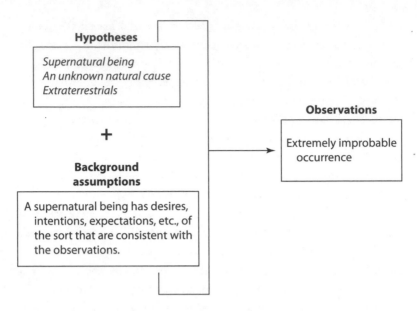

Hypotheses

Supernatural being
An unknown natural cause
Extraterrestrials

+

**Background
assumptions**

A supernatural being has desires,
intentions, expectations, etc., of
the sort that are consistent with
the observations.

Observations

Extremely improbable
occurrence

FIGURE 3.5 Supernatural being versus unknown natural cause versus extraterrestrials

of promoting devotion among Catholics, then she will cause a statue of herself to weep.

With the diagram filled out, we now must ask which of the three hypotheses we should prefer. Which hypothesis does inference to the best explanation justify? As happened in our earlier analysis of Locke's inference, we confront problems immediately. Consider first the background assumption that when a supernatural power wants to promote devotion, it will cause a statue of the Virgin Mary to weep. How do we verify this? Does it just "stand to reason"? I don't see why. Perhaps the reasonable assumption is that true believers don't need to be convinced of the existence of supernatural powers through cheap parlor tricks such as crying statues. The best way to promote faith is rather by other means—for instance, by working in unseen ways to make the world a better, happier, more peaceful place. If *that's* your background assumption, then the observation of a weeping statue points not to a supernatural cause but to something else. And, because verification of either assumption is impossible—we can't simply ask Oprah to sit down with a divine entity for an interview about its goals and methods—we're not justified in believing either of them. Certainly, there's something that doesn't smell right about just deciding to choose whichever background assumption makes your favored hypothesis come out on top.

But similar problems plague the choice among hypotheses. A supernatural being *could* account for the tears flowing from the statue's eyes (granting the background assumptions). However, other hypotheses might explain the tears just as easily. Why not think that a *natural* explanation for the tears is available but beyond our present powers of comprehension? I mentioned in chapter 2 that the Dutch philosopher Baruch Spinoza denied the existence of miracles. He thought that for every purported miracle, there was a natural explanation that we just don't understand. So, for instance, it's easy to imagine that the northern lights I mentioned witnessing were once upon a time considered miraculous by a less scientifically sophisticated culture. Eclipses might once have been regarded as miraculous until the relationships between the earth, moon, and sun were properly understood. If we take as a background assumption the idea that people don't understand everything that goes on in the natural world (and surely this is true!), then the possibility that an unknown natural cause is responsible for the weeping statue is no less compelling than the hypothesis that says something supernatural is the cause.

Finally, why not suppose that extraterrestrials have at their disposal a special beam that makes inanimate objects weep? For whatever reason, they have trained this beam on the statue of Our Lady of Akita, with predictable results. Ridiculous? Certainly. But the evidence in favor of this hypothesis is no worse than the evidence in favor of a supernatural cause. Imagine for a second trying to convince someone who sincerely believes that aliens are responsible for the crying statue that she's wrong, that Mary's spirit is instead the cause. What evidence could you supply that would justify your hypothesis over hers? What could you say that makes your explanation better than hers? If you think the ET hypothesis is unsupportable, you should also think the Mary's spirit hypothesis is as well.

I have explained how to fill in the boxes in order to understand why Locke's inference fails to justify belief in Aaron's miracle and why inference to the best explanation also falls short in justifying the belief that Our Lady of Akita should count as a miracle. I leave it to you to figure out why all the other popular miracles—turning water into wine, parting the Red Sea, healing lepers, raising the dead—similarly cannot be justified by using inference to the best explanation. In each case, the three hypotheses I mentioned in the hypothesis box in figure 3.5 will do an equally good job explaining the observations; in the case of supernatural explanations, essential but unverifiable background assumptions will be required.

A Really Interesting Consequence of These Ruminations

If the discussion so far has kept you on the edge of your seat, what follows is sure to drop you to the floor. Stacked on my desk is a pile of really fat books. In the titles of each are the words *Jesus* and *Resurrection*, among others. The authors of these books, all devout Christians, are tremendously devoted historians or theologians who have spent years learning about the New Testament and the Greco-Roman context in which the events recorded in it purportedly took place. Because I rarely get to use the word *erudite*, I'm going to treat myself to it now. These authors are extremely erudite.

The subject matter of all these books has a fancy name: the historicity of the resurrection. In short, each of these books seeks to make the case for the historical fact of Jesus's resurrection. The common thesis is that the events spelled out in the four gospels regarding Jesus's death, burial, and resurrection report events that actually occurred. Underlying this commitment is the thought that just as historical methods can justify our belief that, say, Caesar crossed the Rubicon in 49 B.C.E. or that Caligula died in 41 C.E., so too historical methods suffice to justify our belief in the resurrection of Jesus.

Opposing this group of scholars is an equally knowledgeable and, since I can't control myself, *erudite* contingent of historians who argue that the historical evidence for Jesus's resurrection is sparse, messy, conflicting, and untrustworthy. Moreover, what evidence there is can be better explained by the assumption that something other than Jesus's resurrection took place—for example, that the witnesses who returned to the tomb and found it empty had in fact gone to the *wrong* tomb (oops!) or that Jesus's body had been removed by grave robbers.

I find this dispute over the evidence for Jesus's resurrection fascinating, and in chapter 6 I summarize the points that I think are most salient in coming to decide whether belief in Jesus's resurrection is justified. However, I wish now to defend a claim that might initially sound quite strange but that follows naturally from the observations I have made about the limitations on inference to the best explanation. The claim is that the dispute over the historicity of the resurrection is pointless. Why pointless? Because although on the surface this debate is about whether Jesus really did rise again, in fact the debate is actually about whether God exists or whether Jesus was God. But Jesus's resurrection can't resolve that issue. Even if we grant that Jesus did indeed rise from the dead, nothing follows about his or anyone else's divinity.

Ironically, it's one of the pro-resurrectionist scholars who convinced me of the irrelevance of the resurrection (though this was surely not his intention). This scholar, Michael Licona, dedicates a few pages in his massive book *The Resurrection of Jesus: A New Historiographical Approach* to responding to a claim made by Bart Ehrman, who has written a number of books questioning the evidence for Jesus's resurrection.[3] Among Ehrman's reasons for skepticism is a limitation he sees in historical methods. Ehrman has argued that the resurrection is a theological issue—if it happened, it was a miracle and as such beyond the capacity of historians to discover. Licona, in contrast, thinks that it should be possible for historians to establish the occurrence of miracles no less than the occurrence of anything else. What historians often *cannot* do, he says, is determine the *causes* of the events whose occurrence they establish: "historians could conclude that Jesus rose from the dead without deciding on a cause for the event. They can answer the *what* (i.e. what happened) without answering the *how* (i.e. how it happened) or *why* (i.e. why it happened)."[4]

But if Licona is right that historians can answer the what questions but not the how or why questions (and I confess that this claim seems strange to me, but Licona's the expert historian here, not I), then he cannot infer from the presumed fact of Jesus's resurrection to his desired conclusion— that a miracle took place. This is because whether something is a miracle, even on Licona's own reckoning, depends very much on the causes involved. An eclipse that happens through natural causes is no miracle. But an eclipse that comes about through God's will certainly is. Likewise, a body brought back to life through a doctor's persistent efforts is not evidence of a miracle, but if God is the agent involved in raising the dead, then we're once again in the realm of miracles. So when Licona asserts that historians cannot answer the how or why questions, he is denying that historians can ever demonstrate that miracles have taken place.

Licona seems oblivious to the fact that he has just deflated his own meticulously researched case for historically proving the miracle of Jesus's resurrection. But why can he not see his mistake? Because he apparently doesn't understand how inference to the best explanation works. Licona casually remarks, "Most would admit that if Jesus rose from the dead, God is probably the best candidate for the cause."[5] Of course, what "most would admit" is not at all a good reason to believe something. There was a time when most people in this country would have admitted that people with dark skin are inferior to those with white skin or that women aren't as smart as men or that cigarettes are harmless. The fact that most people might regard Jesus's resurrection as being caused by God and therefore a miracle tells us absolutely nothing about whether, as Licona assumes, God really is the best candidate to explain the resurrection.

But, if not God, then what? We just have to construct another diagram like the ones I built earlier to see why God is not a better explanation of the resurrection than any number of competitors. Into the observation box goes Jesus's resurrection—remember, I'm simply granting for now that Jesus really did die and then rose to life again. Into the hypothesis box goes God—Licona's choice for the best explanation of Jesus's resurrection. But why stop with just this one hypothesis? In my discussion of Aaron's serpent, I mentioned some other explanations that would serve just as well in the present context to explain Jesus's resurrection. Perhaps seventeen gods worked together to bring Jesus back to life. Or maybe a nearly omnipotent being, not quite as powerful as the Christian god but darn close, brought Jesus back from the dead. Could aliens have restored Jesus's life? That's no worse an hypothesis than *God*. We might also consider the possibility that on very rare occasions—*very very* rare—dead people simply come back to life as the result of natural causes. They may even know before they die that they'll be coming back. Perhaps Jesus was one of those very rare people. Why are these explanations of Jesus's resurrection any worse than Licona's? Why is God the best candidate and not seventeen gods or aliens or unknown natural causes?

And don't forget there's another box that must be filled—the box that contains background assumptions. As in the case of establishing God as the cause of Aaron's serpenty staff or Mary as the cause of tears streaming from the eyes of a statue in Akita, the background assumptions necessary to justify God as the cause of Jesus's resurrection cannot be verified. We can't infer that God caused Jesus's resurrection unless we assume that, for example, Jesus was God's son or that God *wanted* to raise Jesus from the dead or that God hoped to amaze the followers of Jesus or that God believed that some good would come of raising Jesus, or that . . . . You see the point. Until we have some way of verifying any of these assumptions, the inference from Jesus's resurrection to God as its cause, far from being "the best candidate," is in fact incredibly weak.

Miracles are events that are extremely improbable and that have supernatural, usually divine, causes. As I said in the previous chapter, these two features of miracles fit together: events regarded as miracles are so utterly improbable—so unlike anything anyone has ever experienced—that they compel their observers to suspect a supernatural agent at work. Likewise, one might think that the occurrence of supernatural events should be very rare, and so when something supernatural does take place, we're likely to be encountering something unlike anything we have ever seen before.

Identifying causes, supernatural or not, often depends on a particular form of inference. This form of inference, inference to the best explanation, is the scientist's best friend because it justifies belief in things too small to see, such as electrons and germs, or too far away, such as black holes, or too far in the past, such as the formation of the continents. But the value of inference to the best explanation extends beyond scientific practice: detectives rely on it to catch thieves; doctors rely on it to identify diseases. You rely on it to figure out who left the newspaper on the stoop, what's making the noise on the other side of the wall, and where the fire is.

Inference to the best explanation also constitutes the only way (or the best way) to justify belief in the supernatural because the supernatural, like electrons, germs, and black holes, cannot be observed directly. We have to infer its existence from the observations at hand. When we see statues crying, people walking on water, the dead rising, we must infer the cause that lies behind these spectacular events. But the problem with relying on inference to the best explanation in these cases is, first, that the inference to a supernatural cause is successful only given the right background assumptions, and we have no way of verifying the background assumptions that are necessary to infer the presence of supernatural causes. Is the Virgin Mary responsible for causing tears to spring from Our Lady of Akita's eyes? Only if we assume that the Virgin Mary wishes to manifest her presence in that way. If she does not, then the tears are a reason to infer that the Virgin Mary is *not* the cause. What justifies our assumptions about how the Virgin Mary wishes to make herself known? With no answer to this question, our background assumptions reveal themselves to be no more than background *guesses* and so prevent inference to the best explanation from leading us to a justified belief.

But, second, inference to the best explanation requires that one hypothesis does a *better* job explaining the observations than any other. Unfortunately for believers in miracles, hypotheses pointing to causes other than the supernatural will always do at least as well at explaining the observations. Maybe the supernatural existence of the Virgin Mary explains the statue's tears, but the tears can be explained just as well by the possibility of a natural but unknown cause. Indeed, we might want to consider the possibility that extraterrestrials have decided to employ their advanced technology to make the statue cry. Because no observations can distinguish between these various hypotheses, inference to the best explanation can never supply a reason to prefer a supernatural explanation over a natural but unknown one, maybe even one that involves alien activities.

4

JUSTIFYING BELIEF IN
IMPROBABLE EVENTS

IN THE PREVIOUS CHAPTER, I PUT TOGETHER THE FIRST OF my two big arguments against justified belief in miracles. The two arguments correspond to the two characteristics that, I think, distinguish miracles from all other events. The events that are regarded as miraculous—water turning into wine, the dead returning to life—should be extremely improbable, or we would regard them as "business as usual"; and the causes of these events should be supernatural and typically divine. The argument in chapter 3 focused on the second feature of miracles—the problem of identifying a cause as supernatural. Because we can't directly observe supernatural causes, we can have beliefs about them only by inferring their presence from evidence that we *can* observe. We don't see the spirit of the Virgin Mary, but we infer that her spirit is present because we *do* see tears springing from the eyes of a statue of her. But, as I explained, inference to supernatural causes is never justified.

Such inferences involve background assumptions that can't be verified (How do we know that the Virgin Mary will reveal her presence by causing statues to cry?), and they neglect other options that do as least as good a job explaining the evidence as a supernatural explanation would (e.g., there's a natural explanation for the tears of which we are ignorant, or extraterrestrials have trained a weeping beam on the statue).

On their own, these problems suffice to make belief in miracles unjustified. If you think that miracles must include a supernatural origin—and I really see no alternative—then you're stuck. The only way out of this jam that I can imagine requires that the believer offer some way other than inference to the best explanation for identifying supernatural causes. What this way would be, I have no idea. How else to discover something you can't directly observe than by inferring its presence from things that you can observe?

But let's suppose I have missed something. Perhaps I took a wrong step somewhere in my first argument against justified belief in miracles. I don't think I have, but I have been wrong before and about much easier stuff (such as whether the shirt I have chosen goes with the pants I'm wearing). In this chapter and the next two, I present my second argument against belief in miracles. This argument concentrates on the idea that we should regard something as miraculous only if it is extremely improbable (just as we should suspect we're in Portland only if the sky is usually gray).

The argument here differs from the one we just encountered in an important way. When discussing the difficulty facing an inference to supernatural causes, I took for granted that the event in question might actually have occurred. Given that Aaron's staff actually turned into a serpent, what can we say about the cause of the transformation? Given that drops sprang from the eyes of a statue of Mary, what inference should we make regarding the forces responsible? As should now be clear, inference to the best explanation fails to justify the belief that such events have a supernatural origin.

In contrast, I now drop the assumption that the kinds of startling occurrences that many regard as miracles did in fact take place. Why should we believe that they did? On the basis of what evidence is belief in such rarities justified? At the end of chapter 1, I alluded to the difficulties that improbability presents to belief in miracles. Most people who believe in miracles, I said, do so not because they have experienced any on their own but because they have heard about miracles from others. Their belief rests on testimonial evidence. But as we'll soon see, although what we learn through testimony might suffice to justify some beliefs, when the belief is about something as improbable as a miracle, it faces an especially hard burden. Not an impossible burden, but a burden that in the case of the miracles reported so far has not been met. This chapter explains why the improbability of an event raises the bar for testimony. The next two chapters make the case that testimony for miracles fails to reach the bar.

Just How Improbable Are Miracles?

There you are, astride a bar stool listening to Jim go on about his favorite subject, the Karnatakan frog. If what he says is true, that frog certainly does seem to be performing miracles. The things it does are truly spectacular. Of that you're certain. And although you have no trouble believing that some human beings can speak many languages and can maybe, given presently unforeseen scientific breakthroughs, cause lost limbs to regrow and might even at some point in the future be able to resurrect the dead (after all, the very recently dead can be revived with medical interventions in some lucky cases), you are absolutely confident that a frog can't do these things.

No way. If Jim's claims are true, he's reporting something more shocking than anything you have ever witnessed. Nothing comes close. Not the time your schmeared bagel fell to the floor and *landed cream cheese side up*, not the time you ran out of fuel *right in front of a gas station*, not the time you thought you forgot to pack your passport but discovered that you had *never removed it from your briefcase the previous time you traveled*, and not even the time you went fishing and, when you gutted the trout you just caught, *found the ring you had lost when you fished the same stream a year ago*. Every one of these events is somewhat to very improbable, but none of them is as improbable as the things that the Karnatakan frog purportedly does. In each of the former cases, we're seeing lucky coincidences, but nothing more. Coincidences are unlikely—that's what makes them coincidences—and the less likely the more surprising; nevertheless, they're not so improbable as to be contrary to everything you have ever experienced. They make for good stories, sometimes stories that stretch credulity, but we want more from miracles. We want to be able to say that they're not simply coincidences. Miracles must be even more improbable than that.

How improbable must miracles be? I don't have a definite answer to that. I would like to be able to say that a miracle should be a one in a million kind of event or one in a billion or one in a kajillion. The problem is that I don't know how to attach odds to many of the events that people have thought to be miracles. What are the odds that a frog in Karnataka can do all the things Jim says it can do if it's a regular old frog? One in a billion? I'm not sure I even understand what that would mean. Does it mean that if you caught a billion frogs, you should expect that one of them can resurrect dead pets? That hardly makes sense. If the Karnatakan frog is like the frogs I used to catch in Strawbridge Pond when I was a child, then the odds must be more than one in a billion that it can raise dead pets because it was *impossible* for those frogs to raise the dead (although I confess that I never gave them the chance). But if the Karnatakan frog is a special kind of frog, *a miracle-working frog*, then the odds of its doing what Jim says it does are more like one in one! That's exactly what miracle-working frogs are supposed to do.

I suggest that, rather than thinking in terms of odds, we settle on something like this: miracles should be completely, overwhelmingly, awe-inspiringly improbable. They should be the kinds of things that only few people in the world have ever witnessed, so rare are they. They should be so absolutely incredible that they make the mind leap to the conclusion that supernatural forces are at work (an inference, as we saw in the last chapter, that cannot be justified!).

This idea of the utter improbability of miracles seems in keeping with our everyday understanding of them. The miracles that come most easily to mind allegedly occurred during Jesus's lifetime, which makes sense if Jesus were actually capable of performing them, but have any occurred since then? Sure, there have been reports of visions of Mary, of "miraculous" healings after prayer to various saints, and so on, but, to my way of thinking, these "miracles" are small-time compared to raising a dead body that had been buried for a few days, walking on water, and feeding five thousand people with a couple loaves of bread. None of the more recently reported healings, for instance, is absolutely impossible given what we know about the human body. Even if stage-four cancer *almost* always ends in death, almost isn't always. When it doesn't kill, this outcome is perfectly consistent with the fact that stage four cancer is *almost* always lethal. Some people beat the odds. They're just lucky.

But if someone has been dead for a few days or a few weeks and then comes back to life, this has to be more than luck. Similarly, if you awake one morning to discover that the leg you lost to an improvised explosive device while on tour in Iraq is back where it used to be, that would be a miracle. Unlike stage four cancer, which we know will sometimes, albeit rarely, go into remission, lost limbs never grow back, at least not on people. It's a suspicious curiosity, I think, that all the reported miraculous healings over the years are not really inconsistent with what medical science tells us, even if they are extremely unlikely, whereas reports of limbs regenerating overnight, of elderly people becoming young again, of women who have had their ovaries removed bearing children, and so on, remain scarce and unsubstantiated.

We'll leave it at this, vague as it might be: reported sightings of miracles should be the most improbable of events. Miracles should be wonders of wonders.

Testimony

Many, if not most, of our beliefs rest on testimony; and for many, if not most, of these beliefs, testimony is good enough. Anyone with access to the Internet can, with just a few keystrokes, find information about pretty much anything—Napoleon's march on Russia, the half-life of carbon-14, the number of tons of coffee produced annually in Kenya, the signers of the Declaration of Independence, the quadratic equation, the year the Beatles appeared on the *Ed Sullivan Show*, the author of *Wuthering Heights*, and on and on. Moreover, although online encyclopedias occasionally get things wrong, for the most part the information they provide is accurate. If we read that the Beatles first appeared on the *Ed Sullivan Show* in 1964, we can be confident that they really did perform on the *Ed Sullivan Show* in 1964. Moreover, if we do harbor doubts about the reliability of the information, in a matter of seconds we can check it against another, independent online source, finding further support (or lack of support) for our belief.

As these examples make clear, much of what we believe derives not from our own experiences with the world, but from reports by others. I believe the moon exists because I can see it with my own eyes, but my only reason for believing that Pierre, South Dakota, exists is reports by others. These reports can come to me in many different ways. I might meet someone who tells me that she has visited Pierre. Or I might look at a map that was created by people who had passed through Pierre. More likely, the map was created on the basis of other maps, and the creator of the new map was, just as I am, relying on testimony by others when locating Pierre in South Dakota. I might hear of the existence of Pierre when the nightly news covers a railroad tragedy that occurred there. Or I might go online to find a list of state capitals, thereby discovering the existence of Pierre. Testimony can take many forms, and for the most part we are ready to believe what we learn on the basis of such testimony.

Of course, not all testimony is credible. Sometimes we must learn the hard way that a particular individual whom we thought to be a friend can't be trusted. Or we come to suspect that a particular news agency biases its reports in a way that paints inaccurate pictures of current events. A witness might be unreliable through no fault of his own, but unreliable nonetheless. Honest and well-meaning doctors might mislead you about your prognosis because the tests they use rest on bad science or small samples. Given the importance of testimony to our knowledge of the world, we do our best not to listen to people who can't be trusted, not to rely on witnesses with poor eyesight, and not to put too much stock in the unfamiliar or untested.

I don't think I have said anything so far in this chapter that comes as a surprise to you. Here's a summary: much of what we believe about the world comes from the reports of others. Most of the time we can trust these reports, but sometimes we cannot. I agree that these assertions are platitudes, but things are about to become more interesting. From these platitudes, it might be tempting to draw a connection between testimony and our old friend, justified belief: when the testimony you receive is reliable, the beliefs you form on its basis are justified. Appealing as this idea is, however, it is wrong.

A Medical Detour

I'm now going to tell a story that will be crucial for understanding why testimony with respect to miracles has to be more than simply reliable if it is to justify belief in miracles. The story might appear to have little to do with whether we can be justified in believing in miracles, but trust me, it does.

Here's the story. Feeling ill one morning, you call in sick from work and visit your doctor. The doctor checks your blood pressure, takes your temperature, removes the stethoscope from the ice water bath where he keeps it, and places it on your back, asking you to cough a few times. Not finding anything of interest, the doctor sends you to the lab for some blood work. The next morning you receive a call from the doctor, who says the lab results are back and you need to speak to him in person. Immediately.

After seating you in the most comfortable chair in his office, he admits that the news from the lab is not good. You have tested positive for pustulitis, a very rare (and fortunately fictitious) disease that causes noxious pustules to form all over your body and that inevitably kills you unless caught early. The happy news in your case is that the disease was caught early, and with treatment your chances of a full recovery are better than half. The sad news is that the treatment involves a series of painful injections that will cause impotence, blindness, and extremely malodorous flatulence. "So malodorous," he says with a chuckle, "that you really don't have to worry about the impotence." The news stuns you, and you sit in silence for a moment trying to come to terms with the doctor's announcement. "How certain are you?" you finally ask.

"Very," he responds. He then explains that the test is among the most reliable ever developed. "It's 99.9 percent reliable," he says with more enthusiasm than seems appropriate. He then explains that this means that if 1,000 people have pustulitis, it would correctly identify 999 of them, missing only one. Moreover, he continues, of 1,000 people *without* pustulitis, it falsely says of only one of them that he or she has the disease. "So," he concludes, "if you're thinking that maybe you're that lucky one in a thousand who tests positive but doesn't really have the disease, then feel free to skip the treatment."

What would you do?

Base Rates

You really don't want to be treated for pustulitis if you don't have to be. The side effects are—well, enough has been said already. You have asked the doctor about the reliability of the test, and he has told you that it's 99.9 percent reliable. This *seems* to mean that there's only a 1 in 1,000 chance, having tested positive, that you don't have the disease. The news appears grim, but how grim is it really? The question you *haven't* asked the doctor but that you should before making your decision about treatment concerns what's called the *base rate* of the disease. The base rate of the disease is just how common the disease is in the general population. The doctor has told you only that the disease is very rare, but what does that mean? "What does that mean?" you ask your doctor. "How rare?"

The doctor confesses he doesn't know. Fortunately, at that instant your friend Stanley the statistician bursts into the office. Stanley heard from your concerned spouse that you would be visiting your doctor, and because Stanley's mission in life is to expose doctors who commit the notorious *base-rate fallacy*, he has rushed to your aid. "Never fear," Stanley announces, "Stanley is here, and he comes with the news that the base rate for pustulitis in the general population is only 1 in 10 million. And that means," Stanley pauses, perhaps waiting for a drum roll, "that your chance of having pustulitis, given that the test gives a false result 1 in 1,000 times, is only 0.0001. In other words," he concludes, "there's only a 1 in 10,000 chance that you actually have pustulitis."

You look at the doctor, who looks at Stanley, who looks at his neatly manicured fingernails. Can Stanley be right? But what about the lab results? What about the reliability of the test? The doctor said that the test was wrong only 1 in 1,000 times. Doesn't that mean that the chance of having pustulitis is 999 in 1,000 if you test positive? How did Stanley get the much more pleasing result—that there's only a 1 in 10,000 chance that you actually have the disease?

"Explain," you and the doctor say simultaneously.

"Right," says Stanley. "The base-rate fallacy is hard to understand when you try to explain it in terms of chances or probabilities, so let's do it another way." And here's what Stanley goes on to say.

The base rate for pustulitis is 1 in 10 million, which means that in a country with 10 million citizens, only one person is likely to have pustulitis. So the doctor was right that the disease is very rare. It's very rare indeed. So let's imagine a country with exactly 10 million citizens, and let's further suppose that the medical community in this country is very concerned about pustulitis. Perhaps they're worried that the disease will mutate and become more common. Whatever the reason, they decide to test everyone for pustulitis. They use the same very reliable test that the doctor used to test you. So each citizen gives a sample of blood, and each sample of blood is tested. Given the base rate of the disease, chances are good that only one person in this country actually has the disease. Let's call this unlucky soul "Patty."

Remember, even though the test is very reliable, it's not infallible. In fact, as the doctor explained, for every thousand healthy people, the test will falsely "testify" that one of them has the disease even though he or she doesn't. So how many people end up testing positive other than Patty? How many who aren't really sick? Well, if we divide up the 10 million people into groups of 1,000, we'll end up with 10,000 groups. In every one of these groups, one person will test positive for pustulitis who doesn't really have it. Of course, there's also a group, but only one group, which contains Patty, and because chances are very good (999 in 1,000) that Patty will also test positive, one group will probably have two members who test positive: Patty and someone not really sick. So in the end one person will test positive for pustulitis in 9,999 groups of 10,000 people, and two people (Patty and someone else) will test positive for pustulitis in the remaining group. This means that of the 10,001 people who test positive, only Patty really has the disease! "Hence," Stanley finishes, "technically, you have a 1 in 10,001 chance of being sick, so I rounded down."

Stanley is right. The doctor had neglected to think about the base rate of pustulitis, and as a result he misinterpreted the significance of the positive results of your test. He committed the *base-rate fallacy*. It's an easy mistake to make because when assessing the chance that you're genuinely ill, you have to accept two claims that seem to contradict each other. On the one hand, the test for pustulitis is right far more often than it's wrong—of 1,000 healthy people tested, it's right about 999 of them not being sick and wrong about just one of them being sick. On the other hand, given the rarity of pustulitis, the test is wrong far more often than it's right—of the 10 million people tested, it was wrong about who was sick 10,000 times more often than it was right. Weird, but when you're dealing with large odds such as 1 in 10 million, which is the base rate of pustulitis, our intuitions are best not to be trusted.

A nifty way to make Stanley's point even clearer involves a simple two-by-two table (table 4.1). The columns of the table represent whether someone has pustulitis (+ Pustulitis) or not (− Pustulitis); the rows represent whether the test gives a positive result or a negative result. When we fill in the cells of the table, it becomes immediately clear why the test for pustulitis gives so many wrong results.

Let me explain what the numbers in the cells mean. Remember that the base rate for pustulitis is 1 in 10 million. That means that the number in the middle cell, first row, has to be 1 because there is only one person (Patty) in the population of 10 million who both tests positive (+ Test) and actually has pustulitis (+ Pustulitis). It's a very good test in the sense that if you actually have the disease, chances are excellent (999 out of 1,000) that the test will find you. We get a zero in the middle cell, second row, because there's no one in the population who tests negative and has the disease. Patty, recall, is the only person who's really sick.

Now let's think about the numbers in the other column. Because the test says of one person out of every thousand healthy persons that he or she has the disease, when you have a population of 10 million people, the

TABLE 4.1 Results of Test for Pustulitis

	+ PUSTULITIS	− PUSTULITIS
+ TEST	1	10,000
− TEST	0	9,990,000

Note: Total number tested = approximately 10 million.

test will end up giving a positive result to roughly 10,000 people who don't really have the disease (− Pustulitis). That's because there are 10,000 groups of 1,000 people in a population of 10 million. Thus, even a test that goes wrong (indicating that a healthy person is sick) only 1 in 1,000 times will end up making 10,000 errors in this population. That's why the number 10,000 appears in the cell for testing positive but *not* having the disease. This leaves the cell on the bottom right for those who don't have the disease and, consequently, tested negative for the disease. You'll notice that the numbers in the four cells add up to 10,000,001. That's OK—the numbers in the cells are meant as close approximations, not exact figures.

To understand Stanley's conclusion, all we have to do now is examine the first row. This row shows everyone who tests positive for pustulitis. Patty is in this row (in the middle cell), but so are 10,000 people who don't have the disease. In fact, the number of people testing positive who *don't* have the disease is 10,000 times greater than the number who do! And that's why, if you have tested positive for pustulitis, you have pretty much nothing to worry about, *even though the test is right 999 times out of a 1,000 about people who actually have the disease.* The question you should be focused on turns out *not* to be how many people *with the disease* the test is right about, but how often the test is right *when it says that someone has the disease.* It is this second question that requires you to pay attention to the base rate of the disease.

One last thing to notice: if pustulitis weren't so rare, the test would be more trustworthy. Suppose, for instance, that instead of infecting only 1 in 10 million people, the base rate of the disease is quite higher—maybe 1 in 10,000. In that case, the test would still falsely say of 1 person out of 1,000 that he has the disease when he doesn't. But because there are only 10 groups of 1,000 in a population of 10,000 rather than the 10,000 such groups in a population of 10 million, this means that the test would be wrong only ten times as often as it was right. A positive test result would mean that there's a 1 in 10 chance that you're actually sick rather than a 1 in 10,000 chance. And, of course, if the disease were more common still— if it infected 1 in 1,000 people—then our test would be even more trust- worthy. Now if you test positive, there is a 50 percent chance that you're really sick. This is because the test wrongly says of one healthy person out of a thousand that he's sick but will in all likelihood correctly iden- tify the one in a thousand who really is sick. So in a group of 1,000, two people will test positive: one with the disease, one without. When the base rate of the disease is 1 in 1,000, if you test positive, you need to think seriously about treatment.

Back to Testimony

I propose we leave our discussion of pustulitis behind for now but keep in mind the lesson it teaches. The usefulness of a test depends on two fac- tors: how good the test is at correctly identifying some feature of the world, such as a disease, *and* how probable that feature of the world is. We saw the way these factors came together when trying to decide whether to trust the test for pustulitis. Even a very good test—one that goes wrong only one in a thousand times—is useless when the feature of the world it seeks to identify is very rare. In fact, it's worse than use- less, because chances are it will incite panic in those who don't know about the base-rate fallacy. Let's now see what this lesson might teach us about the possibility of having justified beliefs in miracles.

Meet Sally. Sally is ten years old, wears her blond tresses in a pony- tail, favors plaid skirts, and polishes her Mary Janes so assiduously that you can see your reflection in them. You have learned over time that Sally is a very good witness. When Sally tells you about something that hap- pened to her during the day, chances are very good that she's telling the truth. She tells you that her classes were taught by a substitute teacher, and she's right. She reports that her friend Martha wore a hot-pink skirt, and, true enough, when you see Martha later in the day, she's dressed in hot pink. When Sally raced home to tell you about the grasshopper she saw on the sidewalk, you had no reason to doubt this life-changing news. Over time, you have come to count on Sally to give you the scoop. The beliefs you form on the basis of her testimony are justified. Like the test for pustulitis, which is very good at identifying those people who really have the disease, Sally is a very good witness to events that actually happened.

But here's the thing about Sally. She may be reliable, but she's also incredibly boring. Although she regards her running commentary on the world to be fascinating, it's not. Not really. The kinds of events Sally tells you about are completely ordinary. The fancy word for such things is *quotidian*, meaning that they happen every day. If we were to regard Sally as a kind of test, she'd be a test for the occurrence of uninteresting events. When Sally says, "It happened!" she may be right 999 times out of a 1,000, but what she's right about is hardly exciting. Until one day . . .

The doorbell rings, followed by the sound of little fists pounding on your door. There's Sally, who, between heaving breaths, blurts the news that she's spent the previous twelve years captive on an alien spaceship. She explains that because of the wormholes through which the ship passed, she hasn't aged a day, and it will seem to the rest of us that she was never gone. She talks at length about the medical tests the aliens had performed on her, the astounding technology she observed, including things such as large black disks that can make music when they spin beneath a needle and all sorts of other incredible-sounding things.

You have always known Sally to tell the truth. To your knowledge, she has never lied to you, never accidentally misled you, never been anything short of a perfectly reliable source of information. Do you believe her story about the alien abduction?

You shouldn't, even though Sally's hardly ever wrong about all of the ordinary, run-of-the-mill sorts of things that make up the content of her daily reports. Here's how to think about Sally in light of our discussion of pustulitis. The test for pustulitis, we saw, turns out not to be very useful. True, if you really have the disease, then the test will almost certainly give a positive result. But because the probability of the disease is so low, a positive test result is worthless. You can trust a positive result when you actually have the disease, but if you don't know that you have the disease, a positive result can't justify a belief that you have it. Given the utter rarity of pustulitis, the best explanation of a positive test result is not that you actually have the disease, but that the test is wrong.

We also saw how adjusting the base rate of pustulitis changes how much you should trust a positive result. As pustulitis becomes more common, more *quotidian*, you have more reason to believe the test when it tells you that you're sick. Remember, the usefulness of the test depends on two factors—how good a job it does at identifying pustulitis and how probable pustulitis is. If pustulitis is *very* common, then the best explanation for a positive test result may well be that you actually have the disease.

Same goes for Sally. She may be a very reliable witness when it comes to all the boring sorts of events she tends to report, but when she claims to have observed something vastly improbable, such as an alien abduction, her reliability is hardly relevant. The "base rate" for alien abductions, let's agree, must be so incredibly low that even though Sally's usually right about the quotidian stuff, she's almost certainly wrong about the aliens. It's fine to trust Sally, but only when we're not trusting her with much. When she cries "Aliens!" the best explanation for her scream is almost certainly not little green men. More likely, it's something else that's causing her cries. Just as the best explanation for testing positive for pustulitis is that the test has erred, the best explanation for Sally's claims about aliens is that she has erred.

The point is obvious when you think about it. You rarely question what a ten-year-old says when she reports that her friend's birthday is today, but if she tells you that she saw Bigfoot or Richard Nixon or a pink squirrel or *was abducted by aliens*, then you have grounds for doubt. The grounds for doubt needn't arise because you now suspect that the child has all of a sudden become a liar. Chances are the child, if she were trustworthy about mundane sorts of events before, is trustworthy about mundane sorts of events still. The doubt comes not from her drop in reliability but from the sheer unlikelihood of what she's reporting. Given the extreme improbability that Richard Nixon was actually the cause of the child's claim to have seen him—or, put another way, given that something *other* than Richard Nixon was *much more probably* the cause of the child's claim—you should doubt that the child really did see Richard Nixon when she says that she did.

So where did Sally go wrong? Why did she report an alien abduction if the chances that one actually happened are profoundly small? This is my story, and I get to tell it the way I want, but don't worry. Sally didn't lie. Sally would never do such a thing. We're free to consider alternative explanations, and that's what we should always do when we face testimony for an extremely improbable event. Given the very small chance that Sally's testimony is correct—that it was caused by an actual alien abduction—we should ask whether there's a greater chance that something *other* than an alien abduction explains her testimony, just as when testing positive for pustulitis, you should regard having the disease as the least likely explanation of the results. Maybe, for instance, Sally had seen a movie about aliens just before going to bed and spent the night dreaming about Lord Xanthor and his crew of slimy, green alien abductors. So vivid was Sally's dream that she awoke certain that she really had been kidnapped. Or maybe Sally's nasty brother slipped some psychedelic mushrooms in Sally's bedtime tea, and Sally ended up taking a hallucinogenic trip aboard a magic spaceship.

The absolutely crucial point is that when we are faced with testimony about something very improbable, such as an alien abduction, we have to ask ourselves one question: What is more likely—that the event really happened, as the witness reports, or that some other explanation for the testimony is true? Put this way, the question to ask about Sally's tale is: Given Sally's testimony, what's more likely, that Sally really was abducted by aliens or that she is simply confused or remembering details of a vivid dream or is the victim of her mushroom-dispensing brother? On the one hand, granting our assumption that the chance of an alien abduction is absurdly small, the probability that Sally is telling the truth is also vanishingly small. On the other hand, what's the chance that she is just confused or dreamed the whole ordeal? After all, Sally is a bright, if somewhat uninteresting, ten-year-old. My money is on some explanation for her testimony that doesn't require the existence of worm-hole-traveling extraterrestrials. You?

Putting the Base Rate to Work

Before returning to the topic of miracles, let's see how recognition of base rates plays out in other contexts. To do this, I want to sketch in more general terms the reasoning on which I have been relying in the discussion of Sally. My conclusion, that we are not justified in believing that Sally was abducted, depends on a two-step argument. The first step draws attention to the importance of base rates when deciding how much faith to put into testimony. The lesson from the discussion of the test for pustulitis is that the reliability of testimony is not always enough to win our trust. Even very reliable tests are much more likely to be wrong than right when they are testing for something very improbable, such as pustulitis or, in Sally's case, alien abductions.

The second step of the argument builds on the results of the first step. Given the vast improbability of an alien abduction, we're in a position to judge that Sally's testimony is probably false. The chance that she really was abducted given that she says she was is very, very low. But this extremely low chance raises an interesting question: Why would Sally claim to have been abducted by aliens if no such thing really happened? It is in response to this question that we must begin to weigh the probabilities of various alternative explanations for her testimony. We know that her testimony is very likely to be false (step one), so we should ask whether her testimony is more likely to have been caused by confusion or vivid dreams or hallucinogenic mushrooms than by a genuine alien abduction. The chance that she testifies to being abducted by aliens because she *really was* abducted, we suppose, is incredibly tiny. That means we should believe her testimony only if we judge that the chance that she was confused, remembering a dream, or hallucinating is even tinier. But, I think, you would have to be terribly gullible to think that. Little girls with big imaginations are much more likely to have made up a story about alien abduction, for one reason or another, than really to have been abducted. We all know that.

Here's a summary of the reasoning that brings us to doubt Sally.

A person testifies that some event occurs, and we know that the event is very improbable.

Step one: Weigh the probability of the truth of the testimony against the improbability of the event. This is where a consideration of base rates plays a role, and it's at this step that we come to see that even very reliable tests or witnesses are likely to be wrong about what they say when they are reporting about an event with an extremely low base rate.

Step two: Weigh the probability of various alternative explanations for the testimony. Given the results of step one, we must ask ourselves whether there is a more likely explanation for the testimony than the occurrence of the very improbable event. Surely this is true in the case of Sally, where an alien abduction seems a far less likely explanation for her testimony than is confusion, hallucination, or, come to think of it, just about anything.

We are now in a position to apply this two-step reasoning to other examples of testimony about vastly improbable events. Think of all the incredible happenings that get sensationalized in tabloids or on television: someone spotted Bigfoot walking a dog in the woods of Montana; the Loch Ness monster overturned a canoe; a UFO crashed in Roswell, New Mexico; ETs constructed the great pyramids in Egypt. How should we evaluate these claims? Step one: weigh the probability of the alleged event against the reliability of the witness who testifies on its behalf. Each of the events I just mentioned is extremely improbable, which means that even if the witnesses to the events—or the pieces of evidence that are taken to provide positive support for the events—are very reliable, chances are good that they are incorrect. Step two: Given the high likelihood that the testimony is false, we must ask whether some explanation other than the actual occurrence of the event is in fact the cause of the testimony. The discussion of Sally illustrated this point—the probability of alien abduction is so small that we should prefer some other way of explaining her testimony—for example, that she was confused or remembering a dream or had ingested hallucinogenic mushrooms.

Let's now talk about Bigfoot, who seems perennially popular. According to a recent Angus Reid Public Opinion poll, about 20 percent of Canadians and 30 percent of Americans believe that a nine-foot-tall, hairy primate traipses through the northwestern regions of the United States and the territories of Canada.[1] On what is this belief based? Some people claim to have seen Bigfoot. Others claim to have found the characteristic footprints for which the creature is named. This is the testimony that has convinced millions of people of Bigfoot's existence. Of course, if eyewitnesses were *infallible*, then we would have to accept that Bigfoot had in fact been observed even if the existence of Bigfoot is extremely unlikely. Likewise, if the footprints could have been caused by nothing other than Bigfoot, then their presence would be proof positive that Bigfoot is real.

Yet, of course, witnesses are never infallible. Sometimes a witness is in less-than-ideal conditions for seeing accurately. The terrain in which Bigfoot purportedly lives is full of trees that might obscure one's view and of animals such as bears that, when walking upright, might be confused with a bipedal primate. People might also really want to believe in the Bigfoot legend and so are tempted to cry "Bigfoot!" at the sight of almost anything—a raccoon in a tree or an opossum in dim light. And, as we all know, some people like to deceive others. They get a kick out of putting one over on the masses, or they like to make themselves the center of attention. All of these facts explain why no testimony is ever completely trustworthy.

But what about the footprints from which plaster casts have been made or that have been photographed next to human feet to show their astounding size? Most of these photos have been exposed as hoaxes, created from Bigfoot "shoes" worn by mischievous imps with nothing better to do than stomp around in the woods. Lots of the footprints are in fact only partial and could have been caused by just about anything. In short, the footprint "testimony" is far from perfect.

So we have witnesses and evidence that hardly amount to an airtight case for Bigfoot's existence. How reliable are they? What value should we assign to them? Should we say that the witnesses are usually credible, where this means that when reporting on normal, everyday happenings, they are usually correct? Should we say that when witnesses normally report finding footprints from deer or wolves or bears, they are usually right that deer, wolves, and bears are the causes of the footprints?

Fine. Let's say all of those things. What does it mean? As we have seen, we can't know how much stock to put into the credibility of the witnesses and their evidence until we weigh it against the probability that Bigfoot really exists. Just as we can't know whether to believe that someone has pustulitis on the basis of a positive test result until we know the base rate of pustulitis in the population at large, we also can't know whether to believe that Bigfoot exists on the basis of testimony until we know something about the chance that there really is a Bigfoot.

So even before wondering whether to trust the testimony on behalf of Bigfoot, we have to make a decision about the chance that Bigfoot really exists. But how can we know? This is a tough question. Unlike pustulitis, the probability of which can be determined by sampling a large population and counting the number of diseased people, Bigfoot poses an entirely different kind of problem. We might be able to come to a reasonable guess about the probability of Bigfoot by considering certain points. No carcass of a Bigfoot has ever been found. Bigfoot has left behind no teeth or bones that might confirm his or her existence. Think about that for a minute or two. We have found fossil evidence of Cro-Magnons, Neanderthals, Australopithecines, and *Homo erectus*, all of which lived long ago, but never as much as a single fragment from a Bigfoot. If Bigfoots (Bigfeet?) truly exist, they must either be immortal, thus never leaving their remains behind or do something to hide their remains, such as hurling themselves into volcanoes when they decide that they have had enough of their hapless, hirsute existence. Or perhaps they are so entirely rare that we have simply never come across their remains. But if that's true, you should think that you would be even less likely to see a living Bigfoot given that living ones can hide more easily than dead ones.

Also, unless Bigfoot differs extraordinarily from most other animals, we should expect that what's true of birds and bees is also true of Bigfeet: it takes two to make some more. This means that if Bigfoots have been around a long time, there must be a sizable breeding population. It's one thing to suppose that a single Bigfoot might produce only rare sightings, but an entire population? Where's the evidence of such a population?

Of course, sightings of individual Bigfoots raise questions of their own. People have taken photographs of extremely rare creatures living in utterly remote parts of the world, such as albino bats in mountainous caves and two-headed turtles on tiny Indonesian islands. Why has there never been a single clear shot of a Bigfoot? Are they just very camera shy? Do they fear that photographs will steal their souls? Is there something about Bigfoot hair that makes it invisible to cameras? You would think that a hulking nine-foot-tall primate would be easy enough to photograph clearly *just once*.

For all of these reasons, I estimate the chances of Bigfoot's existence to be very low. One in a million, I think, would be generous. But now we have to do our balancing act. We have to put on one side of the scale the very small chance of Bigfoot's existence and on the other side the reliability of the testimony for Bigfoot's existence. Now that the question of Bigfoot's existence as based on testimonial evidence has been properly framed, it's not hard to see why 20 percent of Canadians and 30 percent of U.S. citizens are being irrational. Because of the miniscule probability of Bigfoot's existence, we don't even have to think that the witnesses are in general confused, befuddled, or dishonest in order to doubt their claims. Only a little bit of uncertainty about whether Bigfoot's existence is the best explanation of their testimony is enough to raise serious challenges to their story, just as a test for pustulitis, if incorrect only once in a thousand times, is of no practical benefit. And the same is true of the footprints. The very small chance that Bigfoot actually exists means that we should look for other causes of the footprints. Plenty of causes—eroded bear tracks, for instance or even members of a Bigfoot cult who wear Bigfoot "shoes"—strike me as more likely explanations for the footprints than the possibility that Bigfoot is himself the cause.

By now you should understand why I can't bring myself to believe in things such as Bigfoot or the Loch Ness monster or alien abductions. As much as the possibility of these things adds an element of excitement to our lives, and as thrilling as the stories about them might be, we should remind ourselves of a point I have made repeatedly: you can't make Bigfoots, sea monsters, or aliens exist just by wanting them to. If hope were enough to bring something into existence, I would have a big, fat bowl of ice cream in front of me right now, perhaps with a chocolate truffle or two on top. Rationally deciding whether to believe in these exotic things requires knowing how to weigh the evidence in their favor. But when we do this, we find that the vast improbability of Bigfoot, together with the imperfect reliability of the testimony on his (or her) behalf, renders belief in Bigfoot unjustifiable. Too bad, perhaps, for fans of the big, hairy hulk, but once you accept that you can't hope something into existence, you have to look at the evidence, and the evidence is just not sufficient to justify belief in something as improbable as Bigfoot.

But Should We Ever Believe the Improbable?

My finely tuned schnoz has detected the odor of uncertainty from some of my reading audience. Don't fret—it's not a bad odor, and certainly nothing a little additional deodorant can't hide. I think I know what the problem is, and I want to address it before continuing on to the next chapter, where I turn the points given earlier in this chapter directly toward the issue of miracles. As I have mentioned, the purpose of this book is to convince you that belief in miracles, at least to date, has never been justified, but I don't want to convince you with trickery or sophistry or sleight of hand. I'm doing my best to take the arguments one step at a time so that if you disagree with my conclusions, you will be in a position to put your finger on the precise point in the argument where we part ways. We may be at one of those points right now.

Here's the objection that I expect will have occurred to many of you. Obviously, improbable events, *extremely* improbable events, have occurred in the past. In fact, given how long the earth has been around, it's practically guaranteed that things that happen only once every blue moon or every one hundred blue moons will have happened. Eclipses happen, earthquakes happen, deluges happen, and so too volcanic eruptions. Likewise, given that around 7 billion people inhabit the planet, even the very improbable trait of having one blue eye and one green eye, which happens only in 1 in 10,000 of the population, would be present in 700,000 people. And given the number of birthdays that have been celebrated, it's almost certain that someone, at some time, went to blow out the candles on the cake and ended up accidentally burning the tip of her nose, which left a scar in the shape of a unicorn. And let's not forget all the images of Jesus on grilled-cheese sandwiches.

The point is that *improbable* events are not *impossible* events. They do happen, but the argument about base rates given earlier seems to imply that we are never justified in believing in them on the basis of testimony. If that's true then, I have definitely made a mistake. Improbable events have happened, and we can be justified in believing in them.

This objection is one I have often come across in reading about miracles. The defender of miracles acknowledges something like the reasoning I have explained—I would like to take credit for this type of reasoning, but it's been around in one form or another since at least David Hume's famous attack on miracles in the late eighteenth century[2]— and then complains that the argument shows too much. Here's an example from Michael Licona, whom we met in chapter 3: "we could never conclude that a specific lottery winner actually won, since the probability of anyone, much less a specific person, winning the lottery on a specific day is vastly outweighed by the probability that no one will win."[3] Similarly, the Christian apologist Norman Geisler takes the argument I just made to entail that "we should never believe we have been dealt a perfect bridge hand (though this has happened) since the odds against it are 1,635,013,559,600 to 1!"[4] Licona and Geisler's point, it seems, is that any rule of reasoning that forbids one to conclude that an improbable event has occurred will go wrong in just those cases—and there may be many over great spans of time—where improbable events actually *have* occurred. What's the probability that Janet won the lottery? If a billion tickets have been sold and Janet owns just one of them, then her chance of winning is one in a billion. According to this view, given the very low probability of Janet's winning, we should never believe that she has won, even if she has! And although I'm here taking Geisler's word for it, the odds of being dealt a perfect bridge hand are even smaller than one in a billion, making belief that one is holding such a hand even when one is totally unjustified.

I agree with Licona and Geisler that something's gone wrong if we are *never* justified in believing Janet won the lottery or that you're holding a perfect bridge hand. However, I think their examples are not really apt. For one thing, unlike Sally's abduction by aliens or Jesus's resurrection, we have available to us a number of ways of confirming whether Janet won the lottery or whether the bridge hand you have been dealt is perfect. We could ask Janet to show us her ticket. If she refuses, we can ask the other lottery participants whether any of them have the winning number. Likewise, if you don't believe your own eyes when you stare with incredulity at a perfect bridge hand, you can ask someone else to have a look. You can set the cards aside for an hour, take a nap, and then reexamine them. Such strategies don't apply to Sally and Jesus. Sally continues to insist on her abduction without offering any evidence for it, and no additional evidence is forthcoming. Also, obviously, none of those who testified to Jesus's resurrection is still around to make his or her case, and going back in time to visit Jesus's tomb is impossible.

Another crucial difference is that we have *independent* ways of confirming that Janet has won the lottery or that you were dealt a perfect bridge hand. That is to say, the evidence of these events doesn't depend on a single source. Different people, not communicating with each other, can verify the lottery number or the cards of the bridge hand. This, as we'll see in the chapters to come, is not true of the *gospel* miracles, where news about Jesus's miracles depends on a single source or a small group of individuals sharing a single experience. Independence is important because without it an error in a single source will be repeated again and again. That's why when we doubt that something has happened, we want to check the story in different, independent newspapers or hear it from the mouths of people who have no connection with each other. If every newspaper were to depend on a single reporter, there would be no point in reading more than one.

Of course, Licona and Geisler might insist that I have missed their point. Given the slim odds of anyone winning the lottery and the even slimmer odds of drawing a perfect bridge hand, we should doubt both events. But this claim, too, hardly makes sense. Unlike miracles, which don't *have* to happen, someone eventually has to win a lottery. What's the chance of Janet winning? We have already noted that it's incredibly small. But Janet's chance of winning is no smaller than, in fact it's exactly the same as (if the lottery is fair), anyone else's chance of winning. Given that someone will win the lottery, it's as likely to be Janet as anyone else. Thus, when we hear testimony to the effect that Janet won, we have no special reasons to doubt that she did indeed win. And although a perfect bridge hand may never actually be dealt, it might surprise you to learn that the odds of drawing a perfect bridge hand are *exactly the same* as the odds of being dealt any other hand. The worst possible bridge hand, if such a thing exists, is no more or less likely than the best one. Surely this takes some of the steam out of Geisler's objection. By his reasoning, we should never believe we have been dealt *any* hand!

Some of what I have said in response to Licona and Geisler applies easily to other improbable events. The tsunami that devastated Bangladesh in 2004 was a rare event, but if anyone doubts whether it actually occurred, we have available all sorts of confirming evidence from independent sources—eyewitness accounts from thousands of people, video footage, newspaper reporting, and signs of a recovery still under way. We would be foolish not to believe that such an event, singular as it was, actually occurred.

Eclipses that occurred long ago can be confirmed by the sheer number of reports by ancient historians in different locations or by various depictions of them from around the world. Moreover, notably, our astronomical sciences have also advanced sufficiently to pinpoint when eclipses occurred in the past. Surely an appeal to what we know about how the world works should allow us to confirm or disconfirm the occurrence of vastly improbable events—assuming the events have a natural explanation. Unfortunately for reports of miracles, our best sciences can do nothing to bolster confidence in their truth insofar as such reports purport the occurrence of events that have no natural explanation.

In conclusion, Licona and Geisler's worries about the standards to which I am holding justified belief are worth taking seriously, but I think that such worries can usually be addressed. When hearing testimony about an improbable event, the thing to do is to search for additional sources of independent evidence or to allow our knowledge about how the world works to inform our conclusions. However, a point on which I must insist is that the less probable an event, the less support a fixed amount of testimony provides for it. If Licona and Geisler deny this point, they make a grave mistake, for it is something that can be proven mathematically, as my example of the test for pustulitis suggested. You can play with the numbers to see the point. As you decrease the base rate of the disease, making its presence in the population go from 1 in 10 million to 1 in 100 million to 1 in 1 billion, the test, unless it is improved, becomes less and less useful. As I said, this is a mathematical fact, and I don't think Licona and Geisler would wish to deny a mathematical fact.

One More Objection: How Does the World Work?

I know how defenders of miracles are likely to respond to what I have just said. I know because that stack of books I mentioned in the previous chapter—the books about Jesus's resurrection—have been written by smart and thoughtful people who anticipated some of what I said in my response to Licona's lottery and Geisler's bridge-hand examples. When I mentioned improbable events such as eclipses and tsunamis, I pointed out that for at least two reasons we are better positioned to determine whether these things actually happened than we are to determine whether certain miracles happened. The first is that our evidence for these natural events is better than the evidence for miracles (I develop this point more fully in the next two chapters). We have more independent sources to rely on and more evidence when investigating whether, say, a volcanic eruption destroyed Pompeii than we typically do when investigating miraculous occurrences. But I also noted that we have a pretty good idea by now which sorts of events are possible and which sorts are not. We know a great deal about how the world works. We know eclipses take place, and we know devastating tsunamis occasionally surge across the ocean. We also know, or think we know, that dead bodies don't come back to life, human beings cannot walk on water, wooden or marble statues don't cry, and water cannot be turned into wine.

This is the place where the authors of the books on the resurrection begin to grumble. Many approvingly cite something that C. S. Lewis, best known for his *Narnia* books, once wrote: "We know the experience against [miracles] to be uniform only if we know that all the reports of them are false. And we can know all the reports to be false only if we know already that miracles have never occurred. In fact, we are arguing in a circle."[5] Lewis's reasoning is that if you deny the possibility of miracles (I have not) or even deem them to be incredibly improbable (OK, I have done this), then you are in effect assuming something about the way the world works that you can't assume without begging the question against the possibility of miracles.

The expression I just used, *begging the question*, is philosophical jargon for a particular type of fallacy. Here's an example of the fallacy in action. Suppose I claim to be the smartest person in the world. "I'm the smartest person in the world," I say to you. However, you have your doubts. Accordingly, you ask me a question: "Why should I believe you?" In response, I clear my throat and say, "Because everyone else is dumber." I have just begged your question. That is, I have answered your question simply be repeating my original assertion, although in different words. What it means for me to be the smartest person in the world is for everyone else to be less smart. Thus, if you're doubting that I am the world's smartest person, you're doubting, too, that everyone's dumber than I am, so you'll be unconvinced of my assurances of superior intelligence.

Lewis charges the miracle denier with begging the question. Suppose the miracle denier asserts that miracles are impossible or at least too improbable ever to believe. The miracle believer is now free to ask, "Why do you think that they're impossible or too improbable ever to believe?" The miracle denier had better not at this point answer, "Because I know how the world works, and it works in a way that makes miracles impossible or too improbable ever to believe." If this is how deniers answer the question, then Lewis is correct that they have just begged it. Obviously, the miracle believer disagrees that the "way the world works" is such that miracles are impossible or vastly improbable. If believers believed *that*, then of course they *too* would not believe in miracles. Whether the world works in a way that precludes miracles is precisely the issue between miracle believers and miracle deniers, and this is why deniers can't simply help themselves to a world that works in a way in which miracles don't happen.

Lewis knew what he was talking about. I think many people who doubt miracles are begging the question against believers in just the way Lewis diagnosed. However, I ain't one of these people, for two reasons. First, I have the argument in the previous chapter on my side. How does this argument help? Let's suppose that we, along with the believer, cannot simply assume that the world works in a way that makes miracles impossible. Thinking that everything has a natural explanation, no matter what, begs the question against the believer. Winfried Corduan makes this point nicely in an essay titled "Recognizing a Miracle." There he writes, "It may be possible, by a series of mental gymnastics, to interpret any event in terms of the laws of nature—even if the application of natural laws is not yet fully understood."[6] So Corduan finds fishy the idea that everything that *might* happen in the world can be made to fit into the way we think the world works. *If* (and it's a *big* if, as I show in chapter 6) the evidence justified us in believing that Jesus rose from the dead or that a frog in India speaks several languages, I agree that we would have to do some very serious mental gymnastics to explain these things naturally. Confidence that a natural explanation is *always* available, as Corduan suggests, seems to be cheating.

However, what Corduan says next is a mistake. He concludes, "But there comes a point at which the prima facie presumption has to go the other way—in favor of the miracle explanation."[7] Nope. He's wrong about this. As I showed in the previous chapter, the miracle explanation is, at least given what we currently know, not justified. That's because there exist plenty of other explanations that are just as good as the miracle explanation. For any purported miracle, perhaps God was the cause, but inference to the best explanation gives us no reason to prefer the God explanation to the "almost as powerful as God entity" explanation or the "let's make them think that we're God when we're in fact aliens" explanation. Once these alternatives are on the table, the authors who defend miracles begin to look like the question beggers.

In fact, check this out! In the course of urging open-mindedness on historical investigations of the resurrection, Michael Licona, whom I seem unable not to pick on, says, "What if a god exists who wanted to raise Jesus from the dead? That would be a game changer. In that case, a miracle such as Jesus's resurrection may actually be the most probable explanation. The challenge for historians, of course, is that they cannot know ahead of time whether such a god exists. . . . Historians ought to adopt a position of openness and let the facts speak for themselves."[8]

Well, of course the existence of a god who wanted to resurrect Jesus would increase the probability of Jesus's resurrection. Indeed, if that god were omnipotent, I would say it raises the probability of the resurrection to one. But when Licona goes on to conclude, several hundred pages later, that Jesus was indeed resurrected and that his resurrection should be regarded a miracle, he is violating the very openness that he here claims to value. Why should we regard the resurrection as miraculous unless we assume that God was the cause? What of the possibility that aliens restored Jesus's life or that a nearly but not quite godly entity raised Jesus from the dead? Why should we not take these hypotheses just as seriously as the God hypothesis? In assuming Jesus's resurrection to be a miracle, Licona begs the question against those such as myself who ask, "Why is God the cause and not something else that does just as good a job of explaining the resurrection?"

I said that there are two reasons that Lewis's charge of question begging doesn't stick to me. I just gave you one: I don't assume that there's a way the world works that makes violations of natural laws impossible. Maybe laws of nature can be violated. I doubt it, but I don't insist on it. My point is just that if they are, we're not, presently at any rate, in a position to have justified beliefs about the source of these violations. Maybe God is responsible, but other explanations—explanations *just as good*—are nevertheless available.

Here's the second reason I'm not begging the question when I express doubts about miracles. The argument in this chapter has very limited scope. I haven't tried to show that miracles are impossible. As I have been saying since the beginning of this book, I will never claim that miracles are impossible. I don't think that I can show this, and I wouldn't know how to begin to show it. Moreover, I haven't in this chapter even shown that belief in miracles is not justified (although I did argue for this conclusion in the previous chapter and I helped myself to that argument in responding to Lewis). I have merely presented the *first stage* of a new argument—new in the sense that it is distinct from the one given in the previous chapter that focused just on the nature of inference to the best explanation—that begins with the observation that justified beliefs in improbable events require especially strong evidence. The evidence must suffice to show that an actual miracle, rare as it may be, is in fact the best explanation for the testimony on its behalf. This point is completely consistent with Lewis's claim that we can't rule out miracles by assuming that we know how the world works. I beg no questions in insisting that we need really good evidence to justify our belief in extremely uncommon occurrences. Even if we grant that we don't know how the world works, we should be prepared to concede that *if* miracles can happen, they will be very improbable. It's the improbability of miracles that's significant in the present context. Because miracles are events of extreme rarity, the evidence on their behalf has to be especially good. *No one should take issue with this claim.* Not Lewis, not Licona, not Geisler, not any of the authors of the books about the resurrection that I have been reading.

In closing this chapter, I should confess that the major points I have made do little more than explicate a mathematical result known as Bayes's theorem, which the Reverend Thomas Bayes first articulated in the eighteenth century. The theorem describes how one should update one's beliefs in light of new evidence, so it is perfectly suited to questions about the role of testimony in the justification of belief. In effect, the theorem says exactly what I have been urging in this chapter: if testimony is to justify your belief in an extremely improbable event, you need very good evidence that there's no better explanation for the testimony. Some caution on this point is necessary, however, because the theorem doesn't really speak to the issue of belief or justification. Justification isn't a mathematical idea, and so Bayes's theorem doesn't have anything directly to say about it. However, we can still use Bayes's mathematical insight about how evidence supports a conclusion to guide us in our thinking about justification.

In essence, that's all I have done in this chapter, and by itself it is not enough to call into question beliefs about miracles. Why not? Because perhaps the evidence in favor of miracles, as improbable as they are, is after all sufficient to the task of justifying our belief in their existence. As I have noted, improbable events certainly have occurred in the past, and no doubt we are justified in believing in at least some of these. Next up, then, is stage two of the argument. We now know that evidence for miracles has to be especially strong if our belief in them is to be justified. The evidence must show that the occurrence of an actual miracle, despite its vast improbability, is a better explanation for some bit of testimony than anything else. In the following two chapters, I examine some of this evidence. We will see that it's not especially strong. It's not even *pretty* strong. In fact, it's so weak that one should be worried about putting as much as a pound of weight on it.

5

EVIDENCE FOR MIRACLES

"ENOUGH ALREADY," YOU MUTTER UNDER YOUR BREATH AS Jim drones on about that stupid Karnatakan frog and its miraculous doings. He's relentless.

"And then there was the time that a pet owner showed up with her two-legged poodle," Jim says. "The frog takes one look at the crippled canine, belches out a throaty croak, says something in Cantonese, and just like that," Jim snaps his fingers, "the poodle's back on all fours."

What started as a pleasant evening of chatter with an agreeable stranger has taken a disastrous turn. You're tempted to slide off your stool and be on your way, but by now you've had a beer or three and are feeling a bit surly. Time to ask Jim some hard questions.

"Your evidence for all this froggy business must be really, really strong if you believe it," you start. Jim's moist eyes regain their focus as he trains them on you.

"Why's that?" He asks.

It's the perfect opening to explain to Jim the lessons from the previous chapter. (Funny how Jim asked just the question I wanted him to, isn't it?) The more improbable the reported event, you summarize, the better the evidence must be if our belief that the event actually took place is to be justified. You go through the examples involving an extremely rare disease and the diagnostic test doctors have developed to identify its victims. The rarer the disease, the less you should trust a positive test result. The more likely explanation for a positive test result is *not* that someone has a disease, but that the test result is wrong.

The same point holds of Sally, the extraterrestrials' abductee. When Sally relates the day's normal goings on—schoolyard fights, substitute teachers, magazines in the mail—her usual reliability leaves us with no reason to question the accuracy of her reports. But her claim to have spent twelve years on an alien spacecraft? If we're to believe this, we want more from Sally than her normal reliability. This kind of event is so immensely improbable that even if Sally hardly ever errs, almost *never* reports that something has occurred unless it really has, we should still not believe her. "Hardly ever" and "almost never" don't cut it when testifying to alien abductions. The chance that Sally is mistaken, for whatever reason, is much higher than the chance that little green men had her up to their flying saucer for scones and tea. So, given the extremely small probability that aliens really abducted Sally and the much greater probability that something else explains why she claims to have been abducted, the smart money is *not* on aliens.

To Jim's credit, he seems close to getting it. "Right," he says. "Because the Karnatakan frog's deeds are so incredible, I need pretty good evidence in order to believe it."

"Better than pretty good," you correct him.

"Really, really, good," Jim concedes.

You correct him again: "Better than really, really, good. The evidence should be stronger than just about any other evidence for any other event that you have ever heard about. Remember, endowing a frog with miraculous abilities should be the explanation of *last resort* because just about any other explanation you can think of for the frog's purported deeds will be more probable than that the frog is really a miracle worker."

"Now hold on," Jim puts his palm up to indicate his unhappiness. "Sounds like you're asking for too much. Stronger than any other evidence? What if I see it with my own eyes? That strong enough for you?"

"Doubtful," you say. You then walk Jim through reasoning that should now be familiar. If what the frog seems to do is sufficiently improbable— and, let's admit it, the existence of a frog that speaks several languages, raises the dead, causes limbs to regenerate, and so on is pretty damned improbable—and even if you can trust your eyes 99.99 percent of the time, then the numbers still work out against the frog. You need evidence even stronger than what you get from seeing with your own eyes a frog that seems to perform miracles.

The frown that had been growing on Jim's face blooms fully now. "But why can't I trust my own eyes?" he asks.

How should you know? Nevertheless, you propose some reasons that might explain Jim's lying eyes that are each more likely than the possibility that the frog really is performing miracles. Perhaps, you suggest, the frog lives in a swamp that emits gases that cause hallucinations. Sort of like the Oracle at Delphi who sat atop a crack in the earth from which gases escaped. Her prophecies, in the end, were nothing more than the ramblings of a stoner. Surely hallucination-inducing swamp gases are a more likely explanation of the Karnatakan frog's (apparent) doings than that the frog really has the power to resurrect the dead!

Jim's frown doesn't let up. "You're saying that nothing will ever satisfy you," he whines. "There's no way I can convince you of the frog's miracles if *those* are your standards." Jim looks like he's about to cry into his beer.

You feel for Jim. Your standards are really high. Ridiculously high? That depends. If *ridiculously high* just means really, really, really high, then you're guilty as charged. But if ridiculously high means *unreasonably* high, then you plead innocent. Asking for especially good evidence when told of deeds as singularly improbable as those the frog supposedly performs is nothing if not reasonable. Also, *really, really, really high* is not the same as *impossibly* high. You're not ruling out the possibility of evidence that would justify belief in a miracle-working frog (I'm not sure what kind of evidence this might be, but I fortunately don't have to take a stand on that). You're merely driving home the burden that faces any such evidence. The evidence has to convince you that the reality of the frog's miracles is a *better* explanation than anything else.

An idea occurs to you that might cheer up Jim. You decide to throw him a bone, to cut him a deal. Here's your proposal. You're willing to ease your standards for evidence. In fact, you're willing to ease them dramatically. No longer do you insist that even seeing with your own eyes falls short of adequate. What you ask from Jim is nothing more than the kind of evidence that you would expect to believe something much more likely than a miracle-performing frog. "How about evidence as good as the evidence we have for the moon landing," you suggest, "or Caesar crossing the Rubicon or Lincoln's assassination or even," you finish, "the Miracle on Ice?"

You're being incredibly generous to Jim. Just think about it. Rather than insisting on the kind of evidence that would favor a miracle as the best explanation for some bit of testimony, you're asking for something much, much less. All you want now from Jim is evidence sufficient to convince you of occurrences that, although still relatively uncommon, are nothing more than that. Wars, assassinations, scientific successes, and underdog victories do certainly happen. If Jim can cite evidence for his frog that's only as strong as the evidence that justifies our belief in these sorts of things, you'll be satisfied. A dumber offer has never been made. You would be poor in no time if you were as freewheeling with your finances.

Ground Rules

What should we expect of evidence for miracles? When we hear reports of seas parting, bushes burning, the dead returning to life, statues crying, and water turning into wine, on what grounds should we deem them trustworthy? None of us was present at Lincoln's assassination, yet we believe that he was assassinated, and this belief is justified. Going back even further in time, historians tell us that Caesar crossed the Rubicon. Why should we believe them? When they tell us that their belief about something Caesar did more than two millennia ago is justified, we're right to wonder why. That was a long time ago. What sort of evidence do these historians have that justifies their confidence about something that happened so far in the past?

Before I go further with this question, it's time for me to own up. I'm a philosopher, not an historian. My training is in how to think and reason in a general sort of way. Historical reasoning is a specialized practice that, although no doubt constrained by the same rules of logic that apply to any mode of thinking, also makes use of particular methods and assumptions that go beyond my expertise. Because this chapter, unlike earlier ones, will be assessing historical evidence for various purported miracles, I'm going to be depending much more on the help of others than I have had to so far. Historians, not philosophers, know best about what to look for when trying to reconstruct events of the past, and so the work of historians will figure prominently in this chapter. My aim here is to finish the argument I began in the previous chapter. There we saw why evidence for miracles has to be especially good, and now we will rely on historical considerations to see why that evidence is not good.

The historian Richard Carrier, a skeptic about the miracles reported in the Bible, provides some insight into the kinds of evidence that often satisfy historians.[1] Carrier is responding to Douglas Geivett, who claims that the historical evidence for Jesus's resurrection is just as good as the evidence that Caesar crossed the Rubicon (this claim is common among apologists who wish to read the gospels as history). Geivett goes so far as to claim that evidence for Jesus's resurrection meets "the highest standards of historical inquiry."[2] (Incidentally, Geivett, like me, is a philosopher, not an historian). And, lest one cry foul that I'm relying on a skeptic's conception of good historical evidence, let me just say that, as we'll see, the points Carrier makes about historical evidence are not at all specific to questions regarding the resurrection. They are applicable across the board—that is, to questions regarding any long ago event. You'll see for yourself that they're quite sensible.

One more thing before we start: I can't resist the following philosoph- ical point. There's a quicker reply to Geivett than the one that Carrier makes. Carrier seems to agree, implicitly at least, that if evidence for the resurrection meets the highest standards of historical inquiry, then we should believe that Jesus really did rise from the dead. Carrier's disagree- ment with Geivett is over whether this evidence *in fact* meets these standards. But, as the previous discussion about the Karnatakan frog il- lustrates, we should demand *more* than the highest standards of his- torical inquiry when evaluating evidence of a miracle. Standards of historical inquiry are just not the right standards to apply to testimony on behalf of events as improbable as miracles. If the "highest standards" of historical inquiry demand that evidence for an event be *very* convincing, this still will not do when seeking justified belief in miracles. Because miracles are so much more improbable than run-of-the-mill historical events, we're justified in believing in them only if the evidence for them is *better* than very convincing. Miracles have to be the explanation of last resort, just as you would believe that aliens actually kidnapped Sally only *after* you ruled out every other plausible explanation of her testimony, and you would solemnly accept that you really do have pustulitis only after checking to see that the test didn't err.

Nevertheless, I suggested earlier that we cut Jim some slack, and we might as well extend our generosity to Geivett as well. My reason for men- tioning Geivett in the present context is not, in any case, to examine his particular argument for the historical accuracy of the resurrection. I turn to that issue in the next chapter. We should now make Carrier's response to Geivett our focus instead because it lays out in a convenient list the kinds of evidence that have convinced historians that Caesar did indeed cross the Rubicon. I propose that we use this list as a means for evaluat- ing the various claims about miracles that have been made over the cen- turies. This strategy is at once reasonable and extremely charitable. The reasonableness, we'll see, lies in Carrier's explanation for why histori- ans value the kinds of evidence he discusses; the charitableness rests in the fact that, as I just mentioned, we should by all rights demand much *stronger* evidence for miracles than we do for normal sorts of historical events. Here we go.

Julius Caesar crossed the Rubicon. I'm prepared to say that historians are justified in believing this, and so, too, are we, given the general reliability of the historians from whom we learn facts such as this one. But on what do historians base their belief? What counts for them as good evidence of Caesar's crossing? Carrier discusses five distinct *kinds* of evidence that have convinced historians. I have added descriptive labels to each just to make future discussion of them less cumbersome.

1. Written Records

One reason to believe that Caesar crossed the Rubicon is that Caesar, in collaboration with one of his generals, wrote a book (*The Civil War*) in which he as much as said that he crossed the Rubicon. Moreover, this book became a best seller by ancient standards and is still around today. Although Carrier doesn't say so, I imagine that some of the importance of a written record is that such a thing can retain its reliability as it is reproduced and passed down through the ages much more easily than could an oral accounting of a given event. We all know the dangers involved in oral transmission. By the time my daughter tells me what a friend told a friend told a friend told a friend told her, who knows what the truth is? Famously, the story of Odysseus was preserved orally for many centuries, and we certainly have reason to doubt the truth of its tales of witches, cyclopses, and gray-eyed Athena!

I'm not claiming here that *no* instances of oral transmission should be trusted. The point is simply that written records will *in general* be more reliable than oral ones. Nor am I saying that we should believe everything that has been written down. Indeed, this takes us to the next point.

2. Assent of Enemies

Suppose we had written records of Caesar's dramatic crossing that came only from Caesar and his friends. Crossing the river was Caesar's declaration of war against the Roman Republic— surely a brave and daring (or, for a general of lesser skill than Caesar, foolish) thing to do. If Caesar and his buddies were the only ones to tell of the deed, we might suspect that Caesar was trying to promote himself, that he hoped to secure a place in posterity bigger than what he deserved. For this reason, confirmation by the philosopher and orator Cicero that the crossing occurred comes as especially compelling evidence. Cicero loathed Caesar and was committed to keeping the republic intact. Thus, he had an interest in denigrating rather than glorifying Caesar. That he reported Caesar's crossing would be surprising if it had not really happened.

Moreover, if Caesar hadn't crossed the Rubicon but had disseminated a book in which he claimed to have done just that, we would expect to see written refutations from at least some of the numerous readers of his book. We don't have to imagine that only his enemies would disagree with him. Just imagine if President Nixon had written a book claiming that he was not involved in the Watergate break-in. We would have a bookcase full of refutations. That we have no written refutations of Caesar's claim makes it all the more believable.

3. Physical Evidence

In addition to the written reports of Caesar's crossing, coins exist with images related to the crossing. Moreover, historians have laid their hands on inscriptions that mention battles premised on the crossing as well as records of the conscription of soldiers involved in those battles. If Caesar had lied about crossing the Rubicon, or if for whatever reason the crossing never happened, these bits of physical evidence would be very hard to explain. Why bother to mint coins depicting an event that never occurred?

4. Reliable Accounting

In addition to the written records of Caesar's crossing from sources contemporary with the event, we have the written reports that historians produced a couple hundred years after the event. These historians—most notably Plutarch—have a track record of reliability because they recorded numerous events in addition to Caesar's crossing that we know to be true. But of special importance is that they relied, in many cases, on *independent sources* to confirm what Caesar did on that fateful day in 49 B.C.E. I mentioned the significance of independent sources in the previous chapter. Suppose all of these historians depended on a single source to conclude that Caesar had crossed the Rubicon. If this one source was in error, then so too would all the historians who relied on this source. But if historians could find reports of the crossing from numerous independent sources, then their case quickly gains in strength. Now, if we doubt the historians, we must also accept something very unlikely: that *all* of their sources have for some reason independently made up the same false story! That's like believing of all the students who hand in an identical term paper that they somehow managed to say the same thing with the same words rather than copying them from a single source. Possible, but . . .

We might add a natural corollary to Carrier's claim that we do well to trust historians with a reliable track record: we should not trust reports from people who do not have a reliable track record. Of course, one can fail to be reliable for many reasons. The most obvious reason, I suppose, is dishonesty. However, unreliability can come about for many other reasons. People suffer from biases—often unconscious—that will slant their reports in ways that make them untrustworthy. Superstition and ignorance are other sources of unreliability. Ask yourself this: whom would you sooner trust about the occurrence of some very unusual event a well-educated person who has been trained to think critically and has traveled widely, experiencing many different peoples, places, and things, or a person with little education who is superstitious and largely inexperienced in the ways of the world outside his or her small domain? Obviously you're better off *on average* believing the first person because people of the second sort, as well intentioned as they may be, may not understand what they have seen or what they have heard. They may be too willing to accept stories of strange and fantastic happenings that others with more education and discernment would rightly dismiss. Surely this is why we don't for a second believe a small child who insists that a monster lives under her bed or why we put no stock in the cult member's conviction that his leader can levitate at will. Small children and people under the spell of a charismatic leader are not people in whom we want to invest a great deal of trust.

5. Implicating Consequences

When Caesar crossed the Rubicon, a sequence of events ensued that would not have happened otherwise. One of the greatest nations ever to exist was now at war with itself. The consequences of Caesar's actions were very real and very conspicuous. Indeed, safe to say that had Caesar not declared war on the Roman Republic, we would be living in a world that differs very much from the one in which we live today. I'm not qualified to predict *how* exactly our world would differ—I'll leave that to people who know much more about the historical events that unfolded as a result of Caesar's crossing the Rubicon. However, one reason to believe that Caesar crossed the Rubicon is that we have records of its consequences—the ensuing civil war, the fall of the republic, the institution of an emperor, and so on. All these events that we believe to have happened would not have happened unless Caesar had crossed the Rubicon, so that's just one more reason to believe that he did. They *implicate* the reality of Caesar's crossing.

We now have a list of five kinds of evidence that historians rely on for reconstructing events of the past. Historians will look for written records composed by eyewitnesses or involved parties; they will try to find accounts from not just those who would benefit from telling a story one way rather than another, but from those who have nothing to gain or, even better, might have reasons for wishing the story were false; they will do their best to recover physical evidence of the long ago event's occurrence; they will prefer the testimony of educated, critical, and worldly people with a proven record of reliability to the testimony of well-meaning but ignorant, credulous, and narrowly insulated people; and, finally, they will confirm their beliefs about a past event by tracing the ripples of its consequences.

These five strategies for justifying historical beliefs, I shall say again, should not be dismissed because my source for them happens to be an historian who does not believe in the biblical miracles. Carrier did not sketch out these five strategies for historical research because he "has it in" for miracles. To charge him with making up standards of historical justification only for the purpose of disproving miracles would be unfair. There's no reason to question the wisdom of the list even if, as no doubt a believer such as Geivett would be disappointed to learn, Carrier believes that accounts of the biblical miracles do not measure up.

However, I agree that requiring evidence of *all* five kinds before laying claim to justified beliefs about a particular event in the past *would* be unfair. Nothing Carrier says commits him to the claim that we must have written records, assent of enemies, physical evidence, reliable accounting, *and* implicating consequences to believe that, say, Genghis Khan died in 1227. The Mongols were not known for their literacy, and, as far as I know, the written record available to historians about Genghis Khan's life doesn't amount to much. Likewise, perhaps Genghis killed all his enemies before they had a chance to report his conquests. Nevertheless, we might judge that the physical evidence, reliable accounting, and implicating consequences surrounding his death suffice to justify our belief that he died in 1227. Having all five kinds of evidence would no doubt strengthen our belief about when the khan died, but historians appear satisfied with more limited evidence.

How many kinds of evidence should we require before being confident that a miracle occurred some time long ago? The right answer is this: we ought to expect all five kinds and, if such exist, even more kinds because miracles are so improbable that evidence only good enough to satisfy an historian isn't good enough to justify belief in a miracle. Remember, the occurrence of a miracle should always be the explanation of last resort. The evidence for a miracle such as Jesus's resurrection must be so strong that *any other* explanation for sightings of Jesus after his death would not be as good. However, as I mentioned, I'm willing, for rhetorical purposes, to relax my standards. Forget the right answer. Let's see how well miracles stack up against just the five kinds of evidence Carrier has brought to our attention.

The Book of Mormon

In this section, I show how to apply the criteria for good historical evidence to a particular account of a miraculous happening. And although the miracle I discuss is not one that many people (relatively speaking) believe, it's a good example on which to cut what I hope are your increasingly critical teeth. In the next chapter, we can then use the present discussion as a model for thinking about a much more widely accepted miracle: Jesus's resurrection.

Roughly fourteen million people belong to an organization propounding not only that Jesus was resurrected but also that shortly after his resurrection he toured what are now called the Americas to visit with the descendants of the Israelites whom Mormons believe had settled there. I'll fill in some of the details of his visit soon enough. Of course, if members of this organization, the Church of Jesus Christ of Latter-day Saints, are right about this event, then the thought that we have a bona fide miracle on our hands would be very tempting indeed. After all, the resurrection of Jesus by itself seems to be tremendously improbable and (ignoring my points about limitations on inference to the best explanation) the product of a supernatural cause. Now add to Jesus's resurrection that he then somehow managed to make the schlep across the Atlantic Ocean in 33 or 34 C.E., and we have before us an event that's even more improbable than the resurrection alone. If our belief in a resurrected and America-touring Jesus is to be justified, the evidence for it had better be supergood. But is it even good? Does it meet the standards of historical justification that I outlined earlier? Might there be a better explanation for the reports of Jesus walking the American shores than that he actually did?

Let's now see how well the Jesus-in-America miracle stacks up against the five kinds of evidence that historians often seek. My focus will be on the America part of this miracle rather than on the resurrection part, which we will turn to soon enough. So, specifically, we're now interested in evidence that Jesus came to America.

1. Written Records?

No. If Jesus toured America, he left no written records of his travels, nor did anyone else who might have been around to greet him or travel by his side. Perhaps we should not be surprised. There's no solid evidence that Jesus could write, so we should not expect that he would have produced the first American Baedeker traveler's guide. Moreover, the people to whom he preached were likely not literate. However, just as we're still justified in believing many things about the Mongols despite the fact that they left few or any written records of their accomplishments, maybe we should not be too bothered at the lack of written records of Jesus's American tour.

2. Assent of Enemies?

No. How much more confidence we would have in the story about Jesus if only his detractors had thought to make note of his visit to America! As we saw earlier, our belief about Caesar crossing the Rubicon is stronger thanks to Cicero's commentary. With no love lost between these two men, Cicero would be an unlikely source to tell of Caesar's brave, daring, and calculating move unless he believed he had no other choice, unless he understood that posterity would regard him as foolish for denying or even failing to mention an event so profound. Surely there must have been some who resented Jesus's presence in America or who were angry that he was still alive even after they crucified him. A simple note of disappointment—"Who invited this Jewish guy to our shores?" or "Why did he go all the way over there?"—would be a valuable piece of evidence in favor of the Jesus-in-America story.

3. Physical Evidence?

To evaluate how physical evidence might be used to support belief in Jesus's presence in America, we need some more details about why Latter-day Saints, or, as they also call themselves, Mormons, believe that he did indeed come to America. The full story is complicated and fascinating, but I have space for only parts of it. Mormons believe that the Americas were settled as long ago as 2500 B.C.E. by people fleeing from the Tower of Babel. Because the man who led them to America was named Jared, this group called themselves Jaredites. The Jaredites survived for almost two thousand years but were replaced some years later, in 589 B.C.E., by another group of people, this time from Jerusalem. After a while, that group split into two: the dark-skinned and evil Lamanites and the light-skinned, good Nephites. Over the centuries, the Nephites produced books of scripture that a prophet named Mormon eventually collected into a single volume. Mormon himself wrote an introduction to these scriptures, but not until many years after Jesus supposedly walked among the Nephites. To be precise, Mormon's contribution to what became known as the Book of Mormon purportedly took place in 385 C.E. Mormon's son Moroni completed the book after Mormon's death. Before Moroni died, he buried the book, written on golden plates, in what is now upstate New York.

Fast-forward almost fifteen hundred years. A young man named Joseph Smith (1805–1844), living in Palmyra, New York, claims to have been visited by Moroni, now an angel, during his evening prayers. Smith was a farmer and treasure hunter by trade and so was doubtless quite excited when Moroni told him of the gold plates buried not too far from his home. In no time, Smith had found the location of their burial and recovered the plates along with special eyeglasses for reading them, for the Book of Mormon was written in a variant of ancient Egyptian. Unfortunately, Smith fell short of meeting several requirements Moroni had placed on the recovery of the plates, and as a result it would not be until several years later, in 1832, that Moroni finally allowed Smith to take possession of them.

Once home, Smith set about translating the plates, which required that he wear special spectacles that used seer stones for lenses. Smith eventually traded the glasses for a single seer stone, which he placed into the bottom of a stovepipe hat. He would place his face into the hat and see the translation of the golden plates in the stone's reflection. Translation of the plates apparently did not require actual contact with them. Indeed, Smith would sometimes move about his house, dictating the contents of the plates to a trusted scribe who could hear but not see him.

I'll soon provide more details regarding Smith and the Book of Mormon, but for now we have enough on our plate (not a golden one) to start asking questions about the physical evidence for Jesus's visit to America. If Jesus had really spent time with the Nephites, we should hope for some hard physical evidence. Perhaps a devoted Nephite painted a picture or built a sculpture depicting the encounter. But we have nothing.

Worse, we have no physical evidence that Nephites, Lamanites, or Jaredites ever existed. Mormon scholars debate the population size of these groups, and, as far as I can tell, no consensus on a firm number exists. However, I have seen a range of numbers from about 10 million to 40 million. I'm not sure how to explain that a population as large as perhaps 40 million people could have left no physical evidence of their survival. (Incidentally, the best estimate for the entire world population in 500 B.C.E. is roughly 100 million.) We have physical evidence of small groups of Paleolithic peoples far older than the Jaredites and of bands of hominids older still, but nothing at all of the Egyptian and Hebraic peoples who purportedly crossed the Atlantic starting in 2500 B.C.E. to settle America. Not an arrowhead, not a stone carving, not a bead, not a thing.

But what kinds of things might archaeologists look for? They might start with remains of any of the following, all mentioned in the Book of Mormon, but none believed to have actually existed in the Americas until after Europeans arrived on the scene: elephants, horses, pigs, cattle, goats, sheep, steel, iron, barley, and wheat. With a population in the millions, the Mormon ancestors must have had many times that number of domesticated animals and plants such as those I just mentioned. Why are archaeologists unable to find evidence of any these things prior to the arrival of Europeans? Why is the DNA of Native Americans closer to that of Asians than it is to Egyptians if, as Mormons contend, they are descendants of the nefarious Lamanites?

Moreover, the lack of physical evidence is not for lack of searching. Trusted and neutral organizations such as the Smithsonian Institute and the National Geographic Society have issued statements in which they deny ever finding a single piece of archaeological evidence that supports the claims in the Book of Mormon. The Church of Jesus Christ of Latter-day Saints has itself commissioned archaeological expeditions at the cost of millions of dollars, only to find results no more encouraging than those arrived at by the Smithsonian and the National Geographic Society. Renowned Mesoamerican archaeologist Michael Coe summarizes how the search for physical evidence stands: "As far as I know there is not one professionally trained archaeologist, who is not a Mormon, who sees any scientific justification for believing the historicity of the Book of Mormon, and I would like to state that there are quite a few Mormon archaeologists who join this group."[3]

And what of the gold plates themselves? Smith purportedly revealed them to eleven witnesses, all followers of the religion he created. Some of these witnesses later claimed that they never actually saw the plates but were allowed only to heft the sack in which they were contained. Others denied having contact with the real physical plates, saying instead that they had a vision of them. If the plates did exist, we'll never know because Smith eventually gave them back to the angel Moroni.

This lack of physical evidence should be very troubling for someone who believes that Jesus visited an ancient people in America shortly after his resurrection. Unlike the lack of written records and assent of enemies that might be explained in one way or another, the events chronicled in the Book of Mormon, including Jesus's appearance, should have left physical traces. Naturally, some in the Mormon community have sought to explain away this deficit or have suggested that the Mayan ruins we find today were constructed by ancient Mormon people, despite overwhelming evidence that the ruins in fact date to a later time. Regarding such responses, we must remind ourselves of the argumentative burden. Remember—if Jesus did visit America after his crucifixion, this would be a most extraordinary event. It's very literally incredible, and that's why we think it would be a miracle if true. This means that the evidence for Jesus's visit has to be *better* than simply equivocal or uncertain. The physical evidence should convince us that the *best* explanation for why the Book of Mormon mentions Jesus in America is that Jesus really was in America. I think that even the most committed Mormons should agree that the physical evidence for the events described in Mormon's book is far from definitive. It is certainly not as compelling as the physical evidence we have for Caesar's crossing the Rubicon, and that is all I am very charitably asking for.

4. Reliable Accounting?

Not only did Caesar write about his crossing, but trustworthy historians, relying on independent sources and writing not very long after the event, did as well. In contrast, until Joseph Smith found the golden plates in the early nineteenth century, not a single person over the course of thousands of years thought to mention crossing paths with Jaredites, Nephites, or Lamanites. No histories of the wars between the Nephites and the Lamanites, despite their purported ferocity, exist. Instead, what we know about these ancient races comes from a single source—Joseph Smith. In fact, Smith would have it no other way. He refused to show the golden plates to the many relatives and townspeople who asked to see them. Even those who *might* have witnessed them were never provided with the seer stones that were necessary to understand their contents.

And what of Smith? If Smith were a man of unquestionable character and discernment, these qualities still, strictly speaking, would not be good enough, I think. We're supposed to believe Smith when he says an angel guided him to golden plates that had been buried thousands of years earlier by a race of people who left no physical traces of their existence. This kind of claim, like Sally's claim to have been abducted by aliens, leaves us wanting more than just routine reliability. Smith would have to be nearly infallible if we are to believe that he's telling the truth about the Book of Mormon.

However, what we know of Smith suggests that he may not be a very reliable source—not even as reliable as the historians who routinely report events far less improbable than those that the Book of Mormon describes. Here are some reasons to doubt that Smith really had the golden plates in his possession. At the time and place Smith was raised, people were in the grip of a religious fervor. So prominent were religious revivals during this period that it was called the Second Great Awakening. Smith's home in New York was in the center of it all. In fact, upstate New York was dubbed the *burned-over district* because evangelicals had been so successful in converting the population that there was no "fuel" left to burn. However, Smith's family was not Christian in an ordinary sense. They believed in magic and spoke of having prophetic visions and dreams.

We see the influence of these beliefs in Smith, who would search for treasure by peering into a stovepipe hat in which he kept a "seer stone" (the technique he also used to translate the golden plates) and who also professed to having visions.

These facts about Smith should make a reasonable person question the reliability of his claims regarding the golden plates. Does anyone today sincerely believe that magical stones exist that can be used to detect treasure? Where are these stones? Have they been tested in controlled conditions for their treasure-finding capacities? Smith's acquiescence to a magic that we know today to be false suggests a mind prone to belief in the fantastic. We might also suspect that there's something a little too convenient in the coincidence that an angel should approach a treasure hunter with a story about buried golden plates. For a man as preoccupied with treasure as Smith must have been, golden plates would have come easily to mind. And let's not forget the religious zeal of the Second Great Awakening, of which Smith was a part. Smith lived in a time and place that was infused with religious thought, and he felt its pull no less than anyone. This factor, too, must be included in our judgment of Smith's credibility.

And why did Smith not show his golden plates to the many people who asked to see them—including his own wife? And why, when the person to whom he dictated the contents of the plates lost the first 116 pages, was Smith unable to reconstruct them? If he had read the plates once, could he not have read them again? The questions mount.

All of these points suggest that we have some reasons to doubt Smith's account of the recovery and translation of the golden plates. Even if you don't believe the points to be damning, or if you, like many Mormons, think these points can be addressed in one way or another, I don't think anyone should assert that they are wholly without force. But *they are enough*. Because Smith is recounting a story of extreme improbability, he must be better than a dubious witness; he must be better than merely a pretty good or very good witness. He must be something close to the most impeccable witness ever to have lived. But he's not. In fact, I submit, he's not even as good as the impartial sources that historians prefer, precisely because he was caught up in religious enthusiasm and prone to beliefs in magic that we know to be false.

And, of course, not only does Smith have to be a good witness, but we also have to believe that the *best* explanation for his claims about the golden plates is that they and by implication all that they purport really existed. I think there are many, indeed *thousands*, of better explanations for how we came to have the Book of Mormon than the account Smith provides. For instance, perhaps Smith, wanting to impress his parents with his religious devotion, got caught up in a small fib about seeing an angel. The fib snowballed, becoming bigger and bigger as he tried later to answer questions about it. Before he knew it, his creative and ingenious mind was inventing all sorts of stories, Tom Sawyer–like, rather than "coming clean" with the initial falsehood. The more he talked, the deeper the hole in which he found himself. On this way of seeing things, Smith was a hapless victim of his own imagination.

What's my evidence for this story? I confess not to have any, but surely it is more plausible than golden plates, Tower of Babel refugees, and Jesus's presence in America.

5 Implicating Consequences?

This one's easy. We know Caesar must have crossed the Rubicon because the events that followed would not have happened otherwise. A civil war involving thousands of soldiers would not have taken place had people only *believed* that Caesar crossed the Rubicon. Merely believing he did so would not have transported his army where it had to be.

Compare this now with the Book of Mormon. How would the history of Mormonism have differed if the golden plates never really existed and Smith made the whole story up? To understand the development of Mormonism, it isn't important whether the plates actually existed; all that matters is that people believed that the plates were real and that Joseph Smith had properly translated them. Clearly Smith understood this. Otherwise, he would have been freer with the golden plates. He would have allowed many to see them or at least one other person to translate them.

The fact that the history of modern-day Mormonism can be fully explained on the assumption that the plates did not really exist and that Smith spun their miraculous contents from nothing but whole cloth is the final reason to judge the historical evidence for the miracles of Mormonism to be weak. And though I know I'm repeating myself, I'll close this section by saying once again that even if the evidence were stronger, even if it were as strong as the evidence for Caesar's crossing the Rubicon, it still would not be strong enough to justify our belief that Jesus spent time in America. For something as improbable as that, we should expect something even better than the best historical evidence and certainly better than the zero evidence that we do have.

In this chapter, I presented the second stage of an argument that began in the previous chapter. There we saw why evidence for vastly improbable events, such as miracles, has to be especially good. Evidence for a very uncommon occurrence, if it is to justify your belief in that occurrence, has to be better than the kind of evidence that might justify your belief that a favorite restaurant has closed or that a couple of whom you're fond is breaking up. It must give us reason to favor the occurrence of the uncommon over the more common.

What, then, is the evidence for miracles? Does it meet the standard that would justify us in believing in them? Obviously, an answer to this question depends on a case-by-case investigation of miracles. I undertook the examination of one miracle—Jesus's visit to America—in the previous pages. I argued that evidence for Jesus's American tour doesn't meet even the standards that a historian would demand for having a justified belief in events far less extraordinary. Far from supporting Smith's assertions about the golden plates and their content, all available evidence suggests that Smith's report was false. There's a better explanation for the stories Smith told than the real existence of tribes of Egyptian and Hebraic peoples who settled the Americas and then disappeared without a trace. I suggested such an explanation, and I'm sure you can create one of your own that has a higher probability than that which Smith proposed.

But now that we have seen how to evaluate issues of evidence and justification with respect to one miracle, I would like to spend the following chapter discussing a miracle that is central to the lives of hundreds of millions of people: Jesus's resurrection.

6

JESUS'S RESURRECTION

I HAVE A CONFESSION TO MAKE. IN CHAPTER 5, I COULD have chosen many different purported miracles to illustrate my points regarding the inadequacy of historical evidence to justify belief in miracles. I could have discussed Moses parting the Red Sea, water flowing from a rock, or Aaron's staff turning into a serpent: Where are the written records, the assent of enemies, the physical evidence, the reliable accounting, the implicating consequences? Insofar as each of these events is a miracle, the evidence necessary to justify belief in them ought to be immensely good, but it's not even good by the standards of an historian. I could have discussed any number of other miracles as well, but I instead focused on a miracle recounted in the Book of Mormon. Why?

The truth is, and I mean no offense to Latter-day Saints, most people who first learn of the history and contents the Book of Mormon find it to be, er, ridiculous. Indeed, Matt Stone and Trey Parker, the creators of

South Park, believe Mormonism to be so utterly absurd that it's worthy of a Broadway parody. So I'm betting that many of you will have read my evaluation of Mormonic miracles with something like a conspiratorial attitude. "Well, no kidding," you might have been saying to yourself, "of course that Mormon stuff is a bunch of baloney."

If that's what you're thinking, you're right where I want you because, as I'll now argue, the evidence for a miracle that most Westerners take very seriously, Jesus's resurrection, is not even a teensy bit better than the evidence that Jesus walked the Americas. And, of course, even if it were a teensy bit better, even if it were a great deal better, even if it met the standards that would satisfy an historian, it would still not be enough. That's what all that stuff about base rates in chapter 4 showed. So if you think Mormon beliefs are unjustified, you should also agree that belief in Jesus's resurrection is too.

Kevin

Before tackling the resurrection head on, let's briefly revisit Jim's miraculous amphibian. Suppose that on further interrogation Jim confesses the following. The frog's miraculous deeds happened not recently, but in fact a long time ago. Thousands of years ago. Naturally, he says, people must have witnessed the healings and resurrections the frog performed because how else would we know about them today? However, he admits, not until decades after the frog's miracles occurred did anyone think to write down anything about them. Actually, maybe people wanted to make written records of the frog's actions, he says, but they didn't know how to write. In fact, the people who spread tales of the frog's doings were uneducated and superstitious. They also believed that salamanders, chameleons, and iguanas were capable of similar miracles.

And the authors who eventually did put pen to papyrus to record the frog's miracles? "Don't know who they were," Jim concedes. They lived far away from the frog's home and probably never came in contact with anyone who had seen the frog. Even if they had, they spoke a different language. But, he continues, whoever they may have been, they were utterly convinced of the truth of the stories that had been passed down to them by the superstitious and uneducated people who had heard the stories from other superstitious and uneducated people. In fact, the authors wanted more than anything to spread the good news about the frog so that others, too, would believe. And then, Jim adds, as their texts were translated and copied and transmitted down through the ages, others had a chance to modify them, so no one really knows what the original texts said. But, he assures you, experts who have studied the texts are convinced that many of the changes made to them over the years were done to glorify the majesty of the Karnatakan frog, as the frog no doubt deserves.

Jim's next move surprises you. He reaches down to the foot of his stool and heaves a heavy briefcase up to the bar. Unbuckling the latches, he flips it open and reveals a thick tome with a title embossed in gold running across the center of its cover: *The Life and Times of Kevin the Karnatakan Frog.* "It's all in here," he says, triumphantly. "Every word is true." Should you believe the contents of the book?

Jesus

Before I began researching this book, I had a pretty naive view about the New Testament. Actually, the word *naive* hardly begins to capture my understanding of it. "Profoundly ignorant" is perhaps a better way of putting things. I knew, of course, that the New Testament purports to tell the story of Jesus's life and death and that along the way it exposes us to Jesus's teachings. About that much I was right. But I also thought that the four gospels, Matthew, Mark, Luke, and John, in which Jesus's life is described, were written by four guys named (and here I thought I must be on safe ground) "Matthew," "Mark," "Luke," and "John." These four men, I figured, followed Jesus around, observed his various miracles, listened to him address various groups, discussed his teachings, and so on. Then, I imagined, shortly after Jesus's death they decided they had better write down all they could remember about Jesus so that future generations could be inspired by his life and illuminated by his teachings.

Of course, no one's memories are perfect, but I did expect that the narratives told in the gospels, because they were authored by people who traveled together in Jesus's band of followers, would agree on the main details of Jesus's life and death. There'd be some differences, of course. After all, when my family and I return home from a vacation, friends who ask about how we spent our time might receive slightly different accounts depending on whether they listen to my wife, my kids, or me. We'll emphasize different aspects of the trip. I will focus on what we ate, my wife will talk about the sites we saw, and my kids might talk about the cool hotel rooms. But these differences in what we talk about are not *disagreements*. My wife won't say that we saw the Acropolis, while I steadfastly deny it. That would be weird. The Acropolis is not something you can easily forget. Less-important events, such as whether we had calamari on our first night in Athens or the second night or whether we visited the museum before or after seeing the Acropolis, might generate disagreement, but that's little stuff that hardly matters (and I always seem to be the one who's not remembering correctly).

Anyway, that's what I thought was going on with the New Testament. There might be disagreements between the authors about the little stuff, but surely not the big stuff. Does this view mean I thought that the New Testament could be trusted to justify belief in Jesus's miracles? Not for a second. By now, you know why. Even if the New Testament could lay claim to being a written record of Jesus's life that was composed by people with the intelligence and education that the authors of the gospels obviously possessed, informed by witnesses of the utmost integrity, and even if their accounts were largely consistent, and even if we had all the other kinds of evidence that make us confident in other historical events, such as Caesar's crossing the Rubicon, it still would not justify belief in miracles. Resurrections are not like river crossings. They're far less probable, and so the evidence for them must be far more convincing. If we're to believe that the *best* explanation for the gospel reports of Jesus's resurrection was that Jesus really did rise from the dead, the evidence has to convince us that no other explanation for the reports is more probable. Remember Sally and her tale of alien abduction? The point here is the same: justified belief in incredibly unlikely events demands incredibly good evidence.

Nevertheless, in keeping with the charitable attitude I adopted in the previous chapter, let's see how a New Testament miracle—the BIG one—fares even when we drastically drop our standards for justification. Rather than demanding anything close to the evidence that should actually be required to justify one's belief in Jesus's resurrection, let's ask whether belief in the resurrection is justified even by normal historical standards. My earlier discussion of Mormonism should make familiar the strategy I adopt now, and the most recent update on your conversation with Jim should prepare you for the conclusions.

A last word before I begin: the information I provide in the subsequent sections in this chapter at best scratches the surface of the immense body of literature devoted to the resurrection. I'm not an expert in this literature, but I'm nevertheless confident that I have seen enough to be sure of the one point that matters: the case for the resurrection is far from airtight. Even believers in the resurrection concede that much. But *that concession is enough* given the sheer improbability that Jesus rose from the dead. If the evidence hardly meets the ordinary standards of historical justification, it can't possibly meet the extraordinary standards that justified belief in miracles demands. In the following pages, I do my best to present information that both defenders and deniers of the resurrection accept. Most of what I have learned about the gospels and their history comes from reading historians such as Bart Ehrman and Richard Carrier,[1] both of whom have subjected the New Testament to a level of scrutiny appropriate to the momentousness of the claims it contains. Readers familiar with these authors will recognize my debt to them.

1. Written Records?

No. Jesus, by all accounts, was poor and uneducated. His disciples were too. Chances are good that Jesus and his followers were illiterate. There's some evidence that Jesus could read, but even if he could write, he apparently didn't. If he did, we don't possess anything he wrote, and neither did anyone else. On this historians are agreed. The men who followed Jesus around also couldn't or didn't write anything. They were fishermen, day laborers, and tax collectors. We thus have no written record from Jesus or anyone associated with Jesus of anything Jesus said or did during his lifetime. Nor do we have anything directly from the most likely literate Romans who ruled Galilee at the time—Herod Antipas and Pontius Pilate. And certainly we have nothing that Jesus himself might have written *after* his death.

I mentioned my naive idea that the gospels of Matthew, Mark, Luke, and John were something like transcripts of Jesus's words, written soon after Jesus's death. This, it turns out, is completely wrong, and again I'm not saying anything that you can't easily look up on your own. This is not news (although it was to me). The best evidence suggests that the gospels were written thirty-five to sixty-five years after Jesus's death. We will probably never know who wrote them—the names of the gospels have nothing to do with the names of the authors. In fact, the reason there are *four* gospels rather than some other number apparently owes to the decision of the mid-second-century theologian Irenaeus of Lyons, who took inspiration from his belief that there were four corners of the earth and four principal winds, and so must there also be four gospels. We do know that the authors of the gospels wrote in Greek and that they probably lived far from where the events of Jesus's life transpired. Historians also believe that Mark is the earliest gospel and that the authors of Matthew and Luke based much of what they report on their reading of Mark. John, the latest of the gospels, seems to depend on a different source. Bible scholars believe this because many passages in Matthew and Luke resemble very closely passages in Mark, but passages in John do not. John contains more anti-Semitic passages than the other three "synoptic" gospels and places more emphasis on theological issues and evangelism. I have more to say about the gospels later in this chapter, but the point here is simply that no written records chronicling the events of Jesus's life existed until decades after his death.

2. Assent of Enemies?

Jesus, of course, was not universally adored. Jews wondered at his strange views, and Romans regarded him a troublemaker. You would think that if Jesus returned from the dead, a Roman or Jew might have voiced his disappointment—a simple "Et tu again?" or an exasperated "Oy vey." Just as Cicero felt compelled to record Caesar's crossing of the Rubicon, one might expect that Pontius Pilate or some irritated Jew might have made note of seeing Jesus alive once more. But the historical record of Jesus is largely silent except for the letters of Paul and the gospels, all of which appeared decades after his death and were written by people interested in glorifying Jesus and spreading his message.

In fact, not only do we lack written records of Jesus's death and rebirth, but we also possess no accounts of his life from non-Christians until nearly a century after he was born. Flavius Josephus, a Jewish historian, mentions Jesus twice, but, unfortunately, the writings of Josephus that we possess come to us from Christian translations, so we cannot be sure whether his remarks about Jesus were added to his work by devoted Christians or appeared in the original.

In addition to Josephus, a couple of Romans, Pliny the Younger and Tacitus, also mention Jesus in histories they wrote in the first part of the second century C.E., but only barely, and they dismiss Christianity as a superstition. As evidence of Jesus's resurrection, these mentions count for nothing.

3. Physical Evidence?

Nope. What kind of physical evidence might there be of Jesus's resurrection? We're told that Jesus died and was then buried in a tomb. Three days later a massive stone covering the entrance to the tomb was rolled away, and the tomb was empty. But an empty tomb is an *empty* tomb—it contained nothing that might provide evidence of Jesus's resurrection. It is as much evidence of his resurrection as it is of the work of body snatchers, tomb robbers, or extremely rapid decomposition. (Oh come on! Rapid decomposition? How likely is that? Beats me. How likely is it that a man who's been dead for three days comes back to life?)

Of some interest, however, is archaeological evidence that in Jesus's time the stones used to seal tombs were square and so could not be rolled. If Jesus's tomb was closed with a round stone, whoever did this was at the very vanguard of tomb coverings. He was so far ahead of his time that it wouldn't be until much later—decades after Jesus's death but coincidentally closer to the time that the gospels were actually written—that round tomb coverings were in fashion. It's almost as if the gospel writers were adding details to a story that were left out in the original telling, and because tomb stones were round in their day, they simply assumed that they were also round at the time of Jesus's death. Obviously, this possibility raises questions about whether the authors of the gospels might have been as free with other details of Jesus's life.

4. Reliable Accounting?

We come now to the most substantial evidence for Jesus's resurrection. Many people believe Jesus rose from the dead because they treat the gospel accounts of this event as, well, gospel. Ironically, when one digs into the history of the gospels, the idea that accepting something "as gospel" means accepting it "as certain" flies out the window. Taking something as gospel *ought* to mean "taking something with a grain of salt." Let's see why.

First, as I already noted, the authors of the four gospels never knew Jesus. Given their remoteness from Jerusalem, they probably never knew anyone who knew Jesus. Moreover, the gospels of Matthew and Luke were almost certainly based on Mark's gospel. Thus, unlike historical accounts of Caesar's crossing the Rubicon, which were gathered from independent sources, what we learn of Jesus's resurrection comes to us from at most two sources—that is, the author of Mark and perhaps the author of John, who *may* have based his account on a non-Markan source.

So how did the authors of the gospels gather information about the resurrection? They were apparently not trained historians with the investigative skills and neutrality that one would expect from a reliable source. Almost certainly they were zealous Christians, desirous of spreading their faith, basing their reports on oral recountings that came to them from other zealous Christians, who, in turn, were repeating what they had heard from yet others. This process is not a recipe for reliability. And remember, even if the gospels were to meet standards of historical reliability, that criterion would suffice for justifying belief in ordinary, run-of-the-mill sorts of events, such as deaths, births, wars, elections, and so on. Justifying belief in extraordinary events, such as turning water into wine, feeding thousands with a loaf or two of bread, and returning from the dead, requires evidence of a *much stronger* kind.

One way to investigate the reliability of the resurrection story breaks the issue into two parts. We can first ask about the reliability of the evidence that would have been available to the authors of the gospels. Then we can ask whether the gospel accounts we have available to us today accurately reflect what the gospel authors reported. Analogously, when we consider, for instance, Plutarch's account of Caesar's crossing of the Rubicon, we can ask how Plutarch learned of the crossing. What were Plutarch's sources? What justified his account of the crossing? Then, we can ask whether the copies of Plutarch writings available to us today are true to Plutarch's original texts. In effect, this latter question is about the faithfulness of the processes that led to the Penguin editions of Plutarch that today's interested readers can buy in bookstores (what few bookstores remain, that is). The first question is about Plutarch and his sources: Why should we trust them? The second question is about today's copies of Plutarch's histories: Are they faithful to the originals?

The reason to treat these questions separately should be obvious. Issues of justification can enter in either of two places. On the one hand, perhaps everything Plutarch wrote was true, but the transmission of his originals to the present day involved processes that most likely introduced inaccuracies. For instance, perhaps the scribes who reproduced his histories had available only incomplete copies and so were obliged to invent passages that "seemed" right to them. Or perhaps the scribes had various biases and so felt a need to alter Plutarch's original texts to reflect their own views. In either case, the scribes would have introduced impurities into the original, and the more impurities they tossed in, the less reliable today's copies would be. At some point, we would have to worry that Plutarch's histories no longer justify beliefs about antiquity.

On the other hand, we might suppose that the transmission process from the original Plutarch to today's copies proceeded with complete fidelity. However, this supposed fidelity would not license justified belief in Plutarch's claims unless we also had good reason to believe that Plutarch should be trusted in the first place. A perfectly accurate copy of a false text produces nothing more than another false text.

So let's first ask whether the authors of the gospels should be trusted to tell the truth about Jesus's resurrection. By this question, I don't mean to suggest that the authors were liars. Let's suppose they were not. Rather, I'm simply asking whether what the authors reported was true—whether they had their facts straight. We can next ask whether, even if these authors were likely to have been reliable, the processes of transmission of their original texts would have sustained this reliability.

You might be wondering how we can judge the reliability of the original gospel texts if we have only copies of them. After all, if we have only copies, how do we know what the originals said? And if we don't know this, how do we know whether they were reliable apart from the reliability of the later copies? Good questions! Consider Jim and his frog once more. He now has a text that records all of Kevin the Karnatakan frog's miraculous deeds. Is the text that he has heaved onto the beer-soaked bar trustworthy? I think we can assess the reliability of the original text, which no longer exists, by asking questions about the kinds of beliefs the sources of this text were likely to have had. Were they educated? Sophisticated in the ways of the world? Superstitious? Skeptical? Credulous? All of these questions bear on the general reliability of the content of the text. Just as we shouldn't trust Sally's tale of alien abduction if she is given to flights of fancy or Joseph Smith's account of Jesus in America if Smith was prone to magical and mystical beliefs or if he had special reasons for wanting his story to be true, we shouldn't trust stories of a miraculous frog if they come from sources who too readily misunderstand the ways of the world or who have lurking ulterior motives. In short, understanding something about the minds of the author or authors of a text as well as their sources of information puts us in a position to judge its reliability even if this text is no longer available.

So who were the authors first writing of Jesus's resurrection? On what evidence did they base their narratives? Answering these questions requires that we put ourselves into the sandals of the men and women of the first century C.E. What did these people believe? Would they have made good witnesses? What chance was there that they would report seeing a resurrection even if no such event occurred? I propose that we approach this question from the perspective of another. Ask yourself:

Would you be likely to believe a report of a resurrection today? Would you believe a neighbor who tells you that his sister Sheila died three days ago but showed up for breakfast this morning? I hope that you wouldn't. Why not? Because we have a sophisticated medical science that explains what happens in death and why death is irreversible, except very rarely and certainly not after a period of three days. Moreover, if you believe that Jesus was resurrected, you presumably think his resurrection was a "one off," never to occur again. And if you are the least bit conscientious about seeking justification for your beliefs, you would demand evidence of Sheila's resurrection. You would want an investigation. Perhaps you would request that a team of physicians examine Sheila to determine whether she had in fact been dead. You would want to know whether Sheila had an identical twin who might be posing as the deceased party. As a person who knows that headlines in the *National Enquirer* and other tabloids shouldn't be trusted, you would adopt a healthy degree of skepticism toward any resurrection report. Surely the burden of proof would be placed on those who believe in Sheila's resurrection, and the standards of proof you would demand would be quite high.

Now let's imagine you lived in Roman territory roughly two thousand years ago. Nothing I said earlier about your resurrection scruples would apply. What of medical science? No one could have presented scientific evidence, in anything like the present form, that would explain why dead people remain dead. In any event, those who claimed to have seen Jesus resurrected (and the gospels differ on who these individuals were) knew nothing of medicine. They knew nothing of scientific method. Indeed, historians have amassed a number of oddities that the typical first-century citizen would have believed, none of which, to contemporary eyes, makes a great deal of sense. Had people living back then subjected their beliefs to even the most obvious tests, they surely would have abandoned many of those beliefs. Here are some of my favorite examples of typical first-century Roman beliefs, as collected by Professor Wendy Cotter.[2]

- The saliva of fasting individuals is quite salutary. It can cure eye diseases when rubbed into the eye (as could a mixture of goat dung and honey), and if you are suffering from a pain in the neck, a little saliva dabbed onto the right knee with the right hand and the left knee with the left hand should fix you right up. Saliva is also good for extracting insects that have entered your ear. You merely have to ask someone to spit into your ear, and the critter should come right out. Perhaps most usefully, spitting into your hand after striking someone will soften the victim's resentment.

• Whereas saliva produced during fasting is good for you, beware the flux of menstruating women! This nasty fluid will turn new wine sour, cause fruit to fall from trees, crops to become barren, dull the polish of mirrors, kill bee hives, rust bronze and iron, and drive dogs mad while causing their bite to become poisonous. Naturally, a woman may wish to minimize her monthly flow given its surprising harmfulness. She can do this by placing a *shtikl* of calf gall into her navel. If no calf gall is available, a cup of wine with cucumber juice and water from the ear of an animal will also do nicely.

• Urine, too, has some remarkable properties. Boar's urine in particular is valued for curing earaches. If you are suffering from dropsy, however, a little boar's urine in your drink is just the thing. Even better, though, is dried urine extracted from a dead boar's bladder.

• Does that scorpion look as if it's going to strike you? Saying the word *two* should keep you safe. If a hobgoblin is causing you trouble, heap abuse upon it, and it will run away shrieking.

This advice, keep in mind, was widely accepted as true by people in the first century C.E. One wasn't regarded as superstitious for believing that menstrual fluid could destroy a beehive. The idea that one might test any of these ideas was apparently not part of the local zeitgeist. You might suppose that someone sometime had the idea to apply fasting saliva to one infected eye and not the other to verify its beneficial effects. Or you might wonder why no one bothered to apply menstrual flux to an olive tree to see whether it caused all the fruit to fall (my sister-in-law, who lives in Greece and maintains a small olive grove, would be thrilled to discover this easy alternative to beating her trees with a rake). Apparently, the distinction between superstitious beliefs and justified beliefs was lost on most first-century citizens. *Yet they were the people who would have been relaying stories of Jesus's resurrection to the authors of the gospels.*

"OK," you might be thinking, "people in Jesus's day believed lots of silly things. But what of Jesus's miracles? What of his resurrection? These were singular events. Only Jesus could feed thousands with a loaf of bread. Only Jesus could raise the dead and heal the sick. The singularity of these events means that belief in them was not just 'accepted wisdom,' which, admittedly, might be adopted too uncritically."

I'm not very impressed with this response. The point of bringing attention to the many *obviously* false beliefs of Jesus's contemporaries is to cast doubt on their capacity to understand the world to an extent that would make their testimony about miracles credible. People who believe that hobgoblins could be driven away by heaping abuse upon them display a fantastic gullibility, and not because they're wrong that hobgoblins dislike abuse. Today's four-year-olds would in all likelihood marvel at the nonsense their ancient forbearers believed. But even if one thinks that the singularity of Jesus's miracles speaks in favor of those who reported them, we should pause to wonder how singular they really were. Here are some other beliefs that those living in Jesus's day would have shared:

- Raising the dead is uncommon but certainly not unprecedented. The god Aescleplus raised Hyppolytus from the dead after a sea monster spooked his horses, causing him to be thrown from his chariot. The goddess Isis raised her son Horus from the dead. But mortals as well as gods could raise the dead. Elijah brought back to life the young son of Zarephath's widow almost one thousand years before Jesus raised Lazarus. Then he did the same for the Shunammite woman's son. The philosopher Apollonius, roughly a contemporary of Jesus, resurrected a bride who had died on her wedding day. Among other men who raised the dead we can list Empedocles and Elisha (not to be confused with Elijah).
- Jesus wasn't the only figure to cure leprosy or blindness or paralysis. Elisha was curing lepers a thousand years before Jesus. The Greek king Pyrrhus, who lived three hundred years before Jesus, was a specialist in

healing sick spleens. He would simply sacrifice a white cock and press his right foot against the sufferer's spleen. Apollonius cured a man's limp by rubbing his hands upon the man's hip. He also restored sight to a man whose eyes had been put out and function to a man's paralyzed left hand. The Roman emperor Vespasian restored sight to the blind and revived a man's useless hand.

• Hungry or thirsty? Moses turned sweet the bitter waters of Marah by throwing a branch into them. Elijah provided the widow of Zarephath with a jar of meal that never emptied as well as a bottomless jug of oil. Elisha fed one hundred people with twenty loaves of barley.

Given these facts about the beliefs held by a typical citizen of Jerusalem in Jesus's day, I think we should be very suspicious that the authors of the gospels, who depended on the testimony of such people, were in any position to report correctly those events that transpired immediately after Jesus's death. Just imagine for a minute that you were on a review board that's been charged with authenticating the testimony of one of the gospel writers' sources. I'll call this source Jebbediah. Here's how the transcript of such an inquiry might have looked:

You: Jebbediah, you have been brought before this board for the purpose of ascertaining the truth of your claim that Jesus was resurrected soon after his death.

Jebbediah: Yes, it's true, but I'm having trouble hearing you, so I hope you don't mind if I pour some boar's urine into my ear.

You: By all means. But please report to us the facts surrounding Jesus's resurrection.

Jebbediah: Certainly. It's like this. The day had started badly. I had gone to slaughter a calf so that I could put a schmear of gall into my wife's navel (the flux was strong that morning), but she'd unfortunately bled on my butcher knife, rendering it dull. I hadn't been able to see that the knife had lost its edge because my eyesight ain't so good and I hadn't any fasting saliva available to drip into my eyes.

You: You might have tried goat dung and honey.

Jebbediah: I wish I had thought of that.

You: Please continue.

Jebbediah: Right. So a neighbor tells me that someone said that Jesus's tomb was empty and that someone else might have seen him standing by the entrance.

You: Did that surprise you?

Jebbediah: Not really. Stranger things than that happen all the time. I once got rid of a hobgoblin by calling him names. And haven't you read your Old Testament or Homer? It's all true, and I believe every word. This Jesus feller seems to be following in old Elijah's footsteps. All that healing, and raising from the dead. Elijah went up in a whirlwind, which if you ask me is more impressive.

You: *Old* Testament?

As comic as this inquiry may sound, it may not be far from the truth. The sources for the gospels would have been people who by today's standards would be judged completely unreliable. Any jury of reasonable people would place no trust in such witnesses. But today's standards, informed as they are by two thousand years of intellectual development, are precisely those we should be applying, for we know them to be *better* standards for assessing justification.

In sum, ignorance, credulity, and superstition were pervasive among the first-century C.E. populace. The sources on which the gospel authors would have relied saw the world as a place governed by magic—a place where men could return sight to eyeless sockets, spleens could be healed with the touch of a foot, the dead could be raised. It would be very odd, to say the least, and something much closer to wishful thinking, to believe that although these superstitious people were wrong about almost everything, the one thing they got right was Jesus's resurrection. Why believe they were right *just about that?* Alternatively, you might suggest that they weren't wrong about almost everything. Maybe back then scorpions really would retreat when hearing the word *two*. Maybe raising people from the dead wasn't difficult. But this would mean that Jesus's "miracles" were no more spectacular than Elijah's or Elisha's or Apollonius's or Empedocles's. I don't know why resurrections two thousand years ago would be easier to perform than they are today, but if they were, we would have to wonder why Jesus should be held in higher regard than any of the other people who are reported to have performed similar miracles before and during his time.

So much for the gospel *sources*. Let's now turn to the second issue bearing on justification. Even if the authors of the gospels had got things right—had reported the events surrounding Jesus's death accurately—how do we know that the gospels published in today's editions of the New Testament contain the original text? In fact, we can't know that, and this is because of what we do know. New Testament scholars realized a long time ago that the gospels in their present form would be unrecognizable to their original authors. Nearly two thousand years of errant translations and tendentious redactions have touched almost every aspect of the gospels. Accordingly, it simply makes no sense to wonder whether the gospels can be trusted. What could this mean? *The gospels, in any relevant sense, no longer exist.* New Testament scholar Bart Ehrman has written a number of fascinating books on this topic, and although many Christian Bible scholars reject some of Ehrman's conclusions, few object to his historical claims about the gospels. Indeed, in many cases he is reporting findings that they explicitly endorse.

Ehrman explains in great detail why today's gospels have undergone significant change over the millennia. In large part, this is due to the processes of transcription that transformed the original texts into those we have today. Printing presses did not exist until the fifteenth century, so the earliest copies of the gospels would have been produced by scribes. But, of course, mail delivery did not exist, nor did fax machines or email. These scribes would have been working in isolation from each other, producing copies that would retain whatever errors the editions they copied contained, and they would have added errors of their own. The scribes in the two centuries after Jesus died were untrained in their vocation. Many might in fact have been illiterate, thus unable to detect a misspelling or a missing word. Now fast forward a few hundred years to a time when Christianity had become mainstream and demand for the gospels had increased. Professional scribes were now available to do the copying, but which of the many distinct, locally produced copies of the gospels should they use as a source? The selection would to a large extent have been arbitrary. However, once a selection had been made, the professionalism of the scribes would have ensured some amount of uniformity in later copies. But the originals would already have been lost. Within a century or two of the original drafting of the gospels, no doubt *hundreds* of distinct versions of them already existed.

Things get worse. The earliest copies of the gospels were in Greek. However, the growing demand for Latin copies led Pope Damascus in 382 C.E. to order the scholar Jerome to produce a standard Latin edition. To complete this task, Jerome chose a Greek version from the many distinct versions available to him as well as an early Latin version to create the so-called Vulgate Bible, which would become the standard Bible for Roman Catholics. How he chose from among the myriad Greek texts the one on which he based the Vulgate is anybody's guess. It may have been an especially accurate version of the originals, but maybe not. Of special interest is an apparent discrepancy between the Bible he eventually produced and the Bible that the Dutch scholar Desiderius Erasmus would produce nearly twelve hundred years later.

With the advent of the printing press, Erasmus sought to create a standard Greek edition of the gospels. He chose as his source a single twelfth-century manuscript, which, it turned out, was among the poorer copies available to him. The result, however, became widely used and was the basis of the King James Bible in circulation today. This means that one of the most popular Bibles in existence today derives from a source that was riddled with inaccuracies. Those familiar with the King James Bible might be surprised to learn the following:

• Erasmus found no reference to the Trinity in the Greek source he used for his Bible, so none was included in his first edition. Obviously, this was a big deal to readers of the Latin Vulgate, who found reference to the Trinity in John 5:7–8. What to do? Erasmus refused to insert the passage unless someone could show him a Greek text in which it was included. An enterprising person translated the Latin text into Greek and presented it to Erasmus, who relented. But given that no Greek texts, which are the oldest available to us and thus closest to the originals, mention the Trinity, it seems the author of John never accepted the doctrine of the Trinity.

• The last twelve verses of Mark in which the resurrected Jesus makes an appearance were not included in the oldest Greek manuscripts. Let me repeat that: *the first gospel to have been written did not mention the resurrection*. Mark's account of the resurrection shows up in the King James Bible only because Erasmus chose a Greek translation to which the resurrection had been added at some later date. This means that the gospels of Matthew and Luke, which were based on Mark, added the resurrection story or had it added for them, no doubt by scribes eager to make the text align more closely with their own beliefs.

• "He that is without sin among you, let him first cast a stone at her." Sound familiar? This is from John 8:7. Jesus is chastising the mob that wishes to stone an adulterous woman. The problem, however, is that this passage, like passages about the Trinity and the resurrection, are additions to the Bible made by overzealous scribes. The passage makes no appearances in the earliest copies of the gospels and so likely never was written by the author of John.

What are we to make of these facts? The obvious conclusion will disappoint anyone who thinks that the gospels, as they survive today, are anything like a reliable record of events that happened more than two thousand years ago. Really, how could they be? For most of their history, the gospels were copied by hand by people who saw as their mission the glorification and evangelism of Christian thought. In our own day, we can't even trust certain television news stations, which know their facts will be checked, to get things right. Would Christian devotees who would likely be rewarded for "friendly" amendments to the gospels be any more reliable?

But the problems only get worse. Suppose you simply decide not to worry about the fact that newish copies of the gospels don't resemble the oldest copies. Maybe you're willing to grasp at the hope that the changes over the years have been divinely inspired (Can you justify that belief?) and so discrepancies between old and new can be explained away. Now you face a different question. Which of the four gospels should you believe? As I noted earlier, the gospels don't impart a single narrative. Mark, for instance, never mentions a virgin birth, and neither does John. John mentions the raising of Lazarus and Jesus's turning water into wine, but these miracles don't show up in Mark, Matthew, or Luke. John's is the only gospel in which Jesus is identified as divine. Worse still, variations in the gospels regarding Jesus's resurrection suggest that the authors (or those who added to the texts in the following millennia) twisted their accounts to suit the needs of an audience hungry to believe in the bodily resurrection of Jesus.

Consider, for instance, that the first gospel, Mark, ends simply with an empty tomb. Later additions to this account place three women at the scene, who find the tomb entrance open and a young man dressed in white inside. The boy tells them that Jesus has risen, but there's no sign of him. Chronologically, the next gospel is probably Matthew. Matthew embellishes the tale of Mark (or, given that the tale was not originally in Mark, tells the original resurrection story, or perhaps later additions to Matthew embellish later additions to Mark), adding the occurrence of an earthquake. The boy is now an angel, who appears alongside a bolt of lightning, which paralyzes with fear two guards who had been standing by. The angel instructs the two women who are present (rather than the three in Mark) to spread word of Jesus's resurrection. Then Jesus appears to the women, who grab at his feet, suggesting that his presence is indeed physical. Next comes the tale told by the author of Luke, which begins with three women, not all the same as Mark's three, who find at the tomb not a boy or an angel but two men in dazzling raiment. Again the women are instructed to tell the disciples about Jesus's resurrection. Now, however, when Jesus appears to the disciples, he leaves no doubt of his physical existence. He displays his hands and feet, instructs his disciples to touch him, and then asks for something to eat. Finally comes John. Now just one woman, Mary Magdalene, goes to the tomb. She finds it empty and reports this to Peter and another disciple, who return to the tomb. After they depart, Mary looks into the tomb and finds two angels. Then, turning, she sees Jesus himself, who instructs her to tell the disciples of his resurrection. He leaves nothing to doubt with respect to his physical incarnation when days later he commands doubting Thomas to thrust his hand into the wound at his side.

To anyone reading in chronological order the accounts of the resurrection in the gospels, a nearly irresistible conclusion is that the authors (or the scribes who modified the original texts) sought to clarify to their audience that Jesus was resurrected as a real, physical being. One can almost imagine the author of Matthew saying to himself, "Mark isn't explicit enough that Jesus came back as more than just a vision," and the author of Luke thinking, "Matthew still leaves some room for doubt that Jesus returned as a physical presence." Then the author of John, hoping to put Jesus's physical resurrection beyond all doubt, makes changes to the earlier gospels that would remove any dispute over Jesus's physical incarnation. So who got it right? Was it the author of Mark or the author of Matthew or the author of Luke or the author of John?

5. Implicating Consequences?

Suppose Jesus hadn't really been resurrected. Suppose, instead, people merely believed he had. Why adopt this belief? One reason people have for believing things, as I have mentioned several times, is that they want them to be true. Unfortunately, as I have also said, wanting and hoping something to be true doesn't make it so. Whatever the explanation for why people believe in Jesus's resurrection, the question to ask now is whether Christianity would have developed differently if, on the one hand, the gospels really were true or if, on the other, they were false but people believed they were really true.

One reason for believing that Caesar really crossed the Rubicon is that his army would not be where it needed to be to confront Roman troops had he not crossed the Rubicon. Merely believing that Caesar crossed the Rubicon couldn't explain the ensuing civil war, which involved real soldiers with real swords. Thus, we can point to the consequences of Caesar's crossing as justification for our belief that he really did make the crossing.

But no such parallel is possible for Jesus's resurrection. Whether the gospels got it right and Jesus really was resurrected (unless one thinks the gospel of Mark, in its original form, is the one that got it right) or the gospels got it wrong but people believed them anyway makes no difference to the ensuing development of Christianity. Because mere belief in Jesus's resurrection explains the subsequent history of Christianity every bit as well as the fact of his actual resurrection, nothing justifies a reason to prefer one hypothesis to the other.

This concludes my discussion of the historical evidence in favor of Jesus's resurrection. Keep in mind the larger dialectical point, however. The previous discussion sought mainly to question whether the quality and quantity of evidence we have for Jesus's resurrection would satisfy a historian. I doubt that it could convince an historian who approached the topic with an open mind—not already convinced of the resurrection. If *you* believe in the resurrection, ask yourself this. Suppose you were raised with no religious training. You never heard whisper of Jesus until you were a fully grown adult. Then you encounter for the first time a Christian who provides you with the sort of evidence for Jesus's resurrection that I have described. Would this convince you? Would you be as confident in the resurrection as you are in other events, such as the American Civil War and Kennedy's assassination and the French Revolution? I think honest and reflective persons would have to concede that they would not—or that they *should* not if they place value in having justified beliefs. This reveals the extent to which belief in the resurrection typically doesn't depend on evidence. I would wager instead that most people who accept Jesus's resurrection do so simply because they have been raised to believe it. But, of course, believing all of one's life that something is true doesn't make it true. It doesn't even make it justified.

In short, the weakest claim we can make is that evidence for Jesus's resurrection, unlike the kind of evidence we have for Caesar's crossing of the Rubicon, hasn't created unanimity among historians. Although plenty of historians deny that the evidence suffices to show that Jesus rose from the dead, *no* historian denies that Caesar crossed the Rubicon.

Evidence for the resurrection is nowhere near as complete or convincing as the evidence on which historians rely to justify belief in other historical events such as the destruction of Pompeii or the sinking of the Titanic. No one—believer in Jesus's resurrection or not—can reasonably deny that the evidence for Jesus's resurrection is poorer than it is for Caesar's crossing the Rubicon. It's simply a fact that the wealth of material substantiating an event such as, say, Lincoln's assassination far surpasses what we have in support of Jesus's resurrection. Now, however, we can let the hammer drop. The argument I constructed in chapter 4 comes to the fore. Because miracles are far less probable than routine historical events (volcanic eruptions, sinking ships, assassinations), the evidence necessary to justify beliefs about them must be many times better than that which would justify our beliefs in run-of-the-mill historical events. *But it isn't.* The evidence for Jesus's resurrection is simply not as good as that which historians normally require of events that happen with greater frequency. And even if it were as good, *it is surely not significantly better*, as it must be to justify belief.

Another Perspective

By now my views on the trustworthiness of the resurrection story should be quite clear. In fairness, however, I should emphasize that (no surprise) not all historians of the New Testament would agree with my conclusions. Notably, the eminent scholar N. T. Wright, in his book *The Resurrection of the Son of God*, argues that the New Testament can be read as an historically accurate telling of Jesus's life, death, and rebirth. In fact, according to Wright, once we accept that Jesus's tomb was empty and that many people saw a living Jesus after the discovery of the empty tomb, it is an historical certainty that Jesus must have been resurrected. These two events, Wright thinks, are the *only possible explanation* for why early Christians came to believe in the bodily resurrection of Jesus. Moreover, Wright regards these events as having a "historical probability so high as to be virtually certain, as the death of Augustus in AD 14 or the fall of Jerusalem in AD 70."[3] He continues that "no other explanation" can be given for the rise of Christianity.[4]

Clearly, Wright and I see matters differently. But, I wonder, how can Wright so confidently assert that Jesus's resurrection is an event that we should hold as possessing the highest historical probability—a probability no less than that to which historians assign the date of Caesar's crossing the Rubicon or Augustus's death? And how can Wright be so sure that nothing other than Jesus's resurrection can explain the development of early Christianity?

It seems that Wright's case for the resurrection—consisting of more than seven hundred pages of learned and dense analysis of the historical context in which Jesus and the authors of the New Testament lived—can be easily disassembled with the philosophical tools that I have illustrated in the preceding pages. First and most obviously, notice that Wright's defense of Jesus's resurrection is a conditional one. That is, Wright believes we simply must accept that Jesus rose from the dead *given* that his tomb was found to be empty and that many people saw Jesus alive and well after his crucifixion.[5] But surely one who thinks that we lack justification to believe that Jesus really had been resurrected would doubt that his tomb was in fact empty and would certainly doubt that a number of people saw a living Jesus days after his death. To grant as an accepted fact that Jesus's tomb was empty and that people did see Jesus alive after his crucifixion is to beg the question (there's that fallacy again!). Wright should not be allowed to help himself to the very "facts" that entail the conclusion he wishes to establish. That's just bad reasoning.

But, second, on what basis does Wright believe that the tomb really was empty and that Jesus really did make an appearance to several people? He bases these claims on the accounts in the New Testament. But we just saw that the there is ample evidence working against the New Testament's veracity. The New Testament available to us today likely bears little resemblance to the original collection of writings from which it was constructed. And even if we could establish that it was an accurate copy of the original sources, we have reason to be suspicious of the reliability of these sources, for all the reasons I covered already. Surely, the very least these considerations entitle us to say is that the historical case for Jesus's resurrection is nowhere near as strong as the case for the date of Augustus's death. Wright leaves one with the impression that if we have reason to doubt the historical case for Jesus's resurrection, we might as well also doubt that Lincoln was assassinated or that Louis XVI was beheaded. This conclusion strikes me as entirely unwarranted. And remember, evidence of the sort that's required to establish the occurrence of a miraculous event, in contrast to a "natural" event, must be far stronger than that which might normally satisfy an historian. Thus, even if Wright can convince us of the high *historical* probability of Jesus's resurrection, this effort still falls well short of what he needs to do if he wishes to provide us with adequate justification for believing in the resurrection.

But, finally, let's consider Wright's claim that only Jesus's actual resurrection could explain the growth of Christianity. This is a claim about what I earlier called *implicating consequences*. When seeking to establish that something happened in the past, one might look to the course of events that followed the purported happening. For instance, among the reasons to believe that Caesar really did cross the Rubicon is the fact that Rome found itself in a civil war. Quite possibly, this war would never have happened had Caesar stayed on his side of the river, so the fact that the war did happen is evidence that Caesar crossed the river. Similarly, Wright wishes to claim, Christianity would not have spread and developed as it did unless Jesus had really been resurrected.

But, as I suggested earlier, Jesus's *actual* resurrection is not necessary, as Wright believes, to explain the future course of Christianity. Necessary only was that a group of people *believed* in his resurrection. This belief was all that would be required to spread word of Jesus's rebirth. But, Wright might then ask, what explains this belief? Mustn't we suppose that Jesus really had been resurrected to explain why people came to believe that he had been?

In answering this last question, we must keep in mind the lessons of the many earlier examples we have discussed—Bigfoot, Sally and the aliens, Joseph Smith and the golden plates. Whenever we're presented with a tale of something that is extremely improbable, we need to weigh two probabilities against each other. On the one hand, there's the probability that the event really happened in just the way it was claimed to have happened. On the other hand, for whatever reason, the person reporting the event was mistaken. When weighing these probabilities, one cannot lose sight of the extremely unlikely nature of the event being reported. That's why, for instance, we must give special consideration to Sally's claim that she was abducted by aliens—it's one thing to believe her when she tells us that she saw her best friend, Betsy, earlier in the day, another to believe that aliens took her for a spin in their flying saucer. Because occurrences of alien abduction are extremely improbable, we should believe Sally's account only if we can find no better way to explain why she's telling the story that she is telling. But, of course, there are plenty of better ways to explain why she believes (if she's really sincere) that she was abducted.

Similarly, despite Wright's insistence to the contrary, there are any number of explanations for why early Christians believed that Jesus had been resurrected that do not require Jesus's actual resurrection. And, of course, because Jesus's *actual* resurrection would be an event far more unlikely than any of those mentioned in these other explanations, the *justified* belief is that he was not actually resurrected. What might some of these other explanations be? Here we can let our imaginations run wild and still end up with explanations that are more probable than the one Wright favors. Maybe Jesus had a twin brother, Kanye, whose existence was never known to Jesus's followers. After Jesus's death, Kanye removed his body from the tomb and threw it down a well. He then posed as Jesus and convinced Jesus's followers that Jesus had risen from the dead.

But I can imagine Wright shouting, "*There's no evidence that Jesus had a twin brother! You're just making all that stuff up!*" Well . . . of course I'm making it up. But then, I would ask Wright, which of the following two alternatives strikes you as more probable?

1. The reports in the New Testament should be taken literally, and a human being who was once dead then came back to life. Or
2. Jesus had a twin brother who tricked Jesus's followers into believing that Jesus had returned from the dead.

The first alternative asks us to believe something that is completely beyond what we know to be true about how the world works, whereas the second asks us only to believe that someone had a dishonest twin. Given that both hypotheses explain the reports of an empty tomb and post mortem sightings of Jesus equally well, I contend that we ought to accept the second as being more probable. Why? For precisely the same reason we ought to look for alternatives to Sally's explanation of her recent absence. It could have been an alien abduction, but surely there are more plausible explanations than that—explanations that don't force us to accept the most incredible of events.

Justified belief in miracles requires not just good evidence, not just evidence of a quality that would assure an historian that a war had in fact taken place or a river had been crossed or an election had been rigged or an emperor had been assassinated. Because miracles are so many times more improbable than any of these ordinary sorts of things, justification for believing in them requires evidence that's *super*good. "Supergood" is, I grant, not easily quantified, but whatever it amounts to, it will be better than the evidence that justifies our belief that Caesar crossed the Rubicon or that the Holocaust occurred or that Kennedy was assassinated. Anything short of the best possible evidence will fail to justify our belief in events as improbable as a resurrection.

Do we have immensely strong evidence that Jesus was resurrected? Do we have any evidence that would satisfy an impartial historian? I think that the answers to both of these questions is "no," although a negative answer to the first question is all I really care about.

The people on whom the authors of the gospels depended for their news of Jesus's resurrection had little understanding of how the world works. They were primed to believe all variety of mystical and magical happenings. They can't possibly meet the "supergood" standards that a justified belief in miracles requires when they don't meet more ordinary standards of reliability.

Moreover, the gospels themselves, our best source of evidence for Jesus's resurrection, have undergone so many modifications over the thousands of years since their first drafting that we have no reason for confidence in the reliability of their contents. We simply have no idea what the originals said. We have no idea which of the gospels to believe when they disagree with each other. Did Jesus raise Lazarus or not?

Was he borne of a virgin or not? Was he divine or not? The gospels are far better regarded as artifacts of anthropological or sociological significance than as historical documents. Certainly they fall short of the supergood standard that we must apply to evidence for miracles.

Because evidence for the resurrection fails to meet the supergood standard or even a standard of historical adequacy, we must concede that better explanations exist for the claims made in the gospels. When Sally testifies to her alien abduction, we need to ask whether her testimony is best explained by the occurrence of an actual abduction or whether some other explanation is more probable. Is it more likely that Sally says she was abducted by aliens because she really *was* or because she had seen a movie about alien abductions and then dreamed the whole affair? Similarly, is it more likely that the gospels tell the story of Jesus's resurrection because he really *was* resurrected or because of some other cause? Maybe the eyewitnesses to the alleged resurrection (assuming there really were witnesses!) were superstitious or credulous and were quick to "cry wolf" when in fact no wolf was present. Maybe they had drunk too much wine and simply confused a Jesus look-alike for the real deal. Maybe they reported only feeling Jesus "in spirit," and over the decades their testimony was altered to suggest that they saw Jesus in the flesh. Maybe accounts of the resurrection never appeared in the original gospels and were added in later centuries. Any of these explanations for the gospel descriptions of Jesus's resurrection are far more likely than the possibility that Jesus actually returned to life after being dead for three days. Justification demands that we favor the more probable over the less probable.

7

SHOULD WE CARE THAT BELIEFS

IN MIRACLES ARE UNJUSTIFIED?

THIS LAST CHAPTER IS BRIEF BECAUSE I TAKE MYSELF TO have completed my main task in the previous chapters. I have presented two convincing arguments against justified belief in miracles. After summarizing the main points of the preceding chapters, I want to consider another issue: Should you care if you can't justify your belief in miracles? Should you be bothered if your belief that Moses parted the Red Sea or that Aaron turned his staff into a serpent-eating serpent or that Jesus rose from the dead is no more justified than a belief you form on the basis of an astrologer's prediction or a Chinese cookie's fortune? Should it matter to you that your belief in Jesus's miraculous healing powers is as groundless as Jim's belief in the powers of the Karnatakan frog? Perhaps, you might be thinking, faith doesn't require justification, and faith, after all, is all that's necessary for belief in miracles. Maybe. But maybe not. We'll have to see.

Bye-Bye, Kevin

It's nearly time to bid farewell to Kevin the Karnatakan frog. I have grown attached to this rare *Rana miraculans*, so it's with some sadness that I must confess to have made him up. He doesn't really exist. Surprised? I hope not. Why would anyone ever think he did? What was the source of Jim's confidence that Kevin lounged in the swamps of Karnataka resurrecting dead pets (except for . . .) and regenerating lost limbs, all the while singing to himself in Urdu, Hindi, Navajo, and various romance languages?

Jim's belief in the Karnatakan frog was not justified. Justified beliefs, recall, differ from unjustified beliefs. Justification boosts the probability that a belief is true. When you adorn a belief with justification, you do something to it. You increase the odds that it says something true about the world. Thus, when you add some justification to your belief that the Green Bay Packers will win the Super Bowl, you make your belief more likely to be true. If you wouldn't previously have bet on them winning, now you might be thinking that a wager is in order.

Here's what you don't want to do. You don't want to bet on the Packers simply because you *hope* that they'll win or because you have the conviction that the world would be a much happier place if only they won. The most hopeful hopes and convicted convictions amount to nothing if they don't come with a side order of justification. As I noted earlier, hoping for a chocolate truffle won't get you one. Likewise, as much as Jim might like the idea of a miracle-working frog, and as thoroughly convinced as he is that the world would be a nicer place if only Kevin could be counted on to resurrect pets, these attitudes do nothing to justify belief in Kevin's existence. Without justification, Jim's belief in Kevin does nothing to increase the probability of Kevin's reality.

Naturally, the same points hold for belief in miracles. You may really want it to be true that Jesus rose from the dead, feel completely certain that he rose from the dead, and know with all your heart that the world would be a better place if only Jesus had risen from the dead. I don't doubt that many people feel this way. However, these feelings do nothing to justify a belief in Jesus's resurrection—no more than they do to justify

Jim's belief in Kevin's miraculous doings. If you care about the *truth* of Jesus's resurrection, you need some justification.

So why does Jim's belief lack justification? It fails in two respects, each corresponding to a feature of miracles. Miracles, I have suggested, should be extremely improbable events with supernatural, typically divine, causes. This means that justification for belief in miracles requires being justified in believing *both* that the event in question has a supernatural origin *and* that the event, improbable as it is, actually occurred. Let's remind ourselves of the difficulties that now arise.

First, consider the claim that Kevin's miracles reveal the operations of some divine presence. Maybe, Jim might suggest if pushed, Kevin is an agent of God, if not a god himself, and, just as God worked through Moses to part the Red Sea and through Aaron to turn his staff into a snake, God works through Kevin to resurrect pets, regenerate lost limbs, and so on.

"But why do you believe this?" you ask Jim. "What justifies your belief that God is behind Kevin's miraculous seeming deeds?"

Jim has been reading a book on critical thinking, so he's prepared for the question. "Easy," he says. "It's an inference to the best explanation. People have observed Kevin raising the dead, and the best explanation for how he does this is the presence of a divine force. How else could it happen?"

Jim apparently hasn't arrived at the last chapter of his critical-thinking book. Had he, Jim would have learned that inference to the best explanation justifies a belief only when the observations clearly favor one explanation over another. A divine presence would explain Kevin's abilities, but so too would other things.

"How do you know," you press Jim, "that a god is responsible for all the things Kevin does?" This question might open Jim's eyes a bit wider, for, after all, he has been assuming, "If not a god, what else?"

"What else?" Jim asks.

Now you're free to make up any number of hypotheses that could explain Kevin's feats. The arguments given in chapter 3 come into play here. I like the following hypothesis, which you carefully relate to Jim. Kevin is not native to Karnataka but instead hails from the planet Ranalas, where there exists a species of superpowerful but *not* supernatural beings. These beings look just like good old terrestrial frogs of the sort found in ponds all over Earth. Kevin decided to vacation in Karnataka and while there thought he'd have some fun with the natives, so he set up camp and began his career as a miracle worker. With his advanced understanding of medical science, he raises the dead and regenerates limbs with ease. The languages come with a bit more difficulty, but fortunately he has at his disposal a souped-up version of Google Translate that he can fall back on in a pinch.

Jim looks at you with astonishment. "You're crazy," he mutters under his breath, which by now carries the aroma of hops and malt.

You're crazy? If you accept that the Karnatakan frog can really do everything Jim claims it can, why is the ET explanation any crazier than the God explanation? Both explanations do an equally good job accounting for the observations, so, accordingly, with nothing else to recommend one over the other, Jim's choice is no more justified than yours. Inference to the best explanation, or *better* explanation in this case, fizzles. It can't justify Jim's belief that Kevin is an agent of God or a god himself.

And, of course, what's true of Kevin is true of Moses, true of Aaron, and true of Jesus. Even granting that the people listed here performed the actions for which they receive credit, they cannot be described as *miracle* workers until we are justified in believing that a supernatural force lies behind their deeds. But inference to the best explanation is powerless to adjudicate between a divine hypothesis and one that attributes these figures' powers to, say, extraterrestrials. Maybe Jesus was a superpowerful alien, come to Earth to teach us something about compassion. Is this an obviously worse explanation of how he managed to turn water into wine than one that attributes divine powers to him? I don't see why it would be.

Moreover, don't forget about the problem with what in chapter 3 I called *background* assumptions. Hypotheses explain nothing except in conjunction with other assumptions. Your conclusion that Messy Miranda stole the cookies assumes that messy people leave behind them a pile of crumbs, that cookies crumble, that chocolate melts on fingers, and so on. Each of these assumptions, notice, can be confirmed. For instance, you can break a cookie to test whether it will create crumbs. But the background assumptions to which a miracle believer must appeal are untestable. Aaron's trick with his staff, you might think, must have a divine cause, but why should God want to turn a staff into a serpent? Maybe— and who knows?—all staffs would eventually turn into serpents on their own unless God *prevented* them from doing so. If this were true, then when Aaron's staff turns into a voracious snake, the best explanation would be that God is *not* working through Aaron. Until you can justify your favorite set of background assumptions—the ones that make your favorite hypothesis true—your faith in your hypothesis is just that. Faith.

Of course, the real reason to doubt that Kevin performs miracles has little or nothing to do with whether Kevin is in fact an extraterrestrial. Although Jim doesn't adequately understand the limitations on inferences to the best explanation, the problem of justifying a belief in the genuine cause of Kevin's powers is, at best, an afterthought. What's most troubling about Jim's story is its extreme improbability, which brings us to the second feature of miracles. Let's get serious, as they say. Could Jim's tale really be true? Could a frog with Kevin's talents really exist? Prima facie—that is, at first look—no. All our experience tells against the possibility of such a thing. This is not to say that frogs like Kevin are impossible, but the odds that Kevin actually exists are, well, long. How long? Much, much longer than winning a lottery, I would say, and that's enough to keep any reasonable person from believing in Kevin.

That's the thing about miracles. In addition to being supernatural in origin, or, I suppose, *because* they're supernatural in origin, we ought to expect that they're also tremendously uncommon. Miracles, unlike sunrises, do not happen every day. If they did, we would begin to suspect, as we now do with sunrises, that they have a natural cause. But as I explained in chapter 4, the more improbable an event, the more reliable the evidence on its behalf must be if we're to have justified belief in its occurrence. The example involving the horrible disease pustulitis made this reasoning explicit. Even with an extremely reliable test, one that errs only once in a thousand times, if pustulitis is very rare, then a positive test result fails to justify your belief that you actually have the disease. With an event as improbable as contracting pustulitis, you need more than very good evidence to be justified in believing that you have it.

The lesson for miracle believers is straightforward. If you agree that miracles should be the kinds of event that are utterly uncommon—wonders of wonders—then you must also concede that evidence strong enough to justify your belief in them must be better than merely good. It must be *super*good. This consequence follows from a mathematical theorem. Accordingly, if you wish to argue that your beliefs in miracles are justified, you need to do something very difficult: you need to demonstrate that the evidence for miracles meets this supergood standard, or you need to deny a mathematical fact. I recommend the former strategy.

But what standard must evidence for a miracle meet? I confess that I don't know how to quantify such a standard, but one thing we might say, as I have often pointed out in previous chapters, is that the evidence for miracles should be far better than the evidence an historian would ordinarily need to establish a long-ago event *beyond any reasonable doubt*. In chapters 5 and 6, I considered some miracles, and I examined the historical evidence that apologists have mustered in their favor. I asked about written records in support of miracles, whether dissenters mentioned the miracles, whether physical traces of the miracles remained, whether reliable historians working from independent sources had confirmed the miracles, and whether the occurrence of the purported miracle was necessary to explain the future course of events.

In all respects, the evidence for the miracles I considered fell far short of that which an historian would demand to remove any reasonable doubt about any other historical event. But, of course, if the evidence for miracles can't satisfy even a community of historians, it cannot *possibly* suffice to justify belief in miracles.

So when Jim goes on and on about his favorite topic, he may think his belief in Kevin's doings is justified. But unless he can provide you with evidence far superior to that which would satisfy an historian documenting the history of miracle-working frogs, it is not. You should decline to share his enthusiasm. No one speaking of the frog today has produced any written records from people who had directly observed it. No one suspicious of the frog has written a begrudging concession to its miraculous powers. No physical evidence of the frog exists. People spreading stories about the frog are ignorant and superstitious and conceive of the world as governed by magical and mystical powers. Everything involving the cult surrounding the frog that can be explained on the assumption of the frog's actual existence can also be explained by appeal to mere belief in the frog's existence. Come on, Jim. Really.

With evidence for the frog's deeds so slim, we should start to think that explanations for Jim's beliefs other than Kevin's actual existence might be far more probable. Maybe in his youth Jim spent time in a Kevin cult and was brainwashed. Maybe Jim suffered head trauma from a bicycle accident and awoke from a coma with a certainty in the existence of a miracle-working frog in India. Maybe Jim is especially gullible and was tricked by a nefarious con man who goes door to door selling copies of *The Life and Times of Kevin the Karnatakan Frog*. Anyone with any understanding of how the world works should find these latter explanations of Jim's belief to be vastly more probable than what the belief purports—that there really *is* a miracle-working frog in the Karnataka region of India.

Apply these lessons about Kevin to any miracle of your choosing. That's my case against justified belief in miracles.

So What?

At this point, you might be wondering, "So what?" Maybe you're saying to yourself, "OK, my belief in miracles is not justified. What consequences should this lack of justification have on my life?"

This is a good question. In fact, it's a great question. As I mentioned at the outset of this project, none of my arguments disproves the possibility of miracles. I am focused only on the issue of justification. But whether a belief is justified is independent of the truth of the belief. Justification will raise the probability that a belief is true, but it doesn't make a belief true. Thus, for instance, military experts might be justified in believing that a particular country possesses weapons of mass destruction. If their belief is justified, it is more likely to be true than if it is based on merely a fear that the country possesses the weapons or if it is based on the ramblings of a fortune-teller. We should realize, however, that despite its justification the belief may be wrong. We might end up taking military action against a country on the basis of a justified belief only to find that we were wrong. No sign of weapons of mass destruction anywhere. Conversely, our belief that a country possesses weapons of mass destruction may be *unjustified* but *true*. Perhaps we heard about the weapons from an unreliable source who just happened this one time to be correct.

So miracles may have actually occurred despite everything I have said about whether you are justified in believing in them. I hope I have convinced you that you are not justified in believing in miracles, but my *epistemological* commitments prevent me from saying anymore. I have no arguments up my sleeves against the possibility of miracles.

Phew. That's good news. If I haven't disproved miracles, you might be thinking, no harm in continuing to believe in their existence. It's not like continuing to believe in witches or ghosts, which we know don't exist (humor me if you think they do). So long as belief in miracles doesn't involve believing in anything incoherent (as perhaps Baruch Spinoza believed), why not believe?

I am now going to argue for a claim that sounds very peculiar: given that belief in miracles is not justifiable, you should continue to believe in miracles only if you don't really care much about them—only if, that is, their possibility makes little difference to your life. This idea can be clarified with another story.

Let's imagine that you find yourself once again at your doctor's office. Nothing good ever comes from such visits, as you now realize. Sure enough, your doctor tells you that the test results are in and, sadly, you have pancreatic cancer. Pancreatic cancer, of course, is a terrible disease. It killed my grandmother in short order and often leads to death even when caught early. No doubt, the news devastates you, as well it should.

Now let's tell a happier story. You visit your doctor because you have had a sore throat. She examines you, rubs a swab on the back of your throat, puts it into a culture, and calls you the next morning with the news that you have strep. She issues you a prescription for an antibiotic, wishes you a speedy recovery, and hangs up.

Let's now think about the *significance* of two different beliefs. The first belief is that you have pancreatic cancer; the second, that you have strep.

The first belief, I submit, is more significant than the second. By this, I mean that the first belief will have a greater impact on how you live your life, what decisions you will make, how you will spend your time, and so on than the second belief. For instance, if I believed that I had pancreatic cancer, I'm pretty sure I would quit my job. Understanding that I had only months left to live, I would decide to spend my final months cramming into them all those activities that until now I had always thought I'd like to get around to doing someday. I would also ask my wife to take a leave of absence from her job so that she could be by my side as we toured Europe, dined at fancy French restaurants, and hiked Alaska (not really—my wife and I are not really the outdoorsy types). I might pull my kids from school so that they could join us during my last days. These are BIG life changes.

But upon learning that I had strep, I would most likely not change my life much at all. In fact, I can say I would not with a great degree of confidence because I have contracted strep several times in my life. In each case, I treated myself with antibiotics, but I certainly didn't quit my job or book tickets for a world tour with my wife. I don't remember if I even bothered to tell my kids about my condition. In fact, I'm fairly sure I continued to go to work and, probably, much to my wife's disapproval, maintained my running schedule.

So a belief that you have pancreatic cancer is much more *significant* than a belief that you have strep throat. You ought to care much more about the former belief than about the latter because its consequences are so completely *life changing*.

Understanding this will help us to make sense of the strange claim I made earlier. The more significant a belief in your life, the more you should seek justification for it, and the less significant, the less you should care about having justification for it. Given the radical change in your life a diagnosis of pancreatic cancer would bring, I think you would be foolish not to assure yourself that the test result was accurate. Imagine quitting your job and spending your life savings on a romp around the world only to find yourself, two years later, in perfect health. How disappointing (sort of)! Before pulling your kids out of school and asking your spouse to take a leave of absence from her career, you should seek justification for your belief that you have pancreatic cancer. The significance of the belief demands that much of a reasonable person.

Strep throat, in contrast, is not such a big deal. If your doctor's diagnosis was wrong—perhaps you were the one in a thousand whom the test falsely says has the disease when in fact you don't—so what? The only cost of the bad diagnosis in this case is a few dollars for the antibiotics. Because your belief that you have strep is of little significance, you should invest little time seeking to justify it.

Ask yourself this: If you could choose to justify only one of two beliefs, would it be the belief that you have pancreatic cancer or the belief that you have strep throat? You see the point.

Time now to turn our attention, one last time, to Jim. Jim's belief in Kevin the Karnatakan frog lacks justification for all the reasons I mentioned earlier. Should Jim care? Should he continue to believe in the frog despite the absence of justification? This depends, I think, on the significance of the belief in Jim's life. Suppose that, despite all of Jim's drunken fascination with the frog, in his more sober moments (which, for Jim, are the norm), Jim rarely thinks about the frog. He goes weeks without once wondering what Kevin's been up to. He reads books to his children about a certain frog and a certain toad without associating them with Kevin. Now and then, however, he takes a bus home from work and finds himself seated next to a Froggy—that is, a member of the Kevin cult. He notices the sign of Kevin—a picture of a frog wearing a turban and sitting on a lily pad—on the passenger's baseball cap. Then he thinks to himself, "That Kevin. He's something else all right." But that's it. After the passing thought, Jim goes back to studying the *People* magazine he hides from his wife.

If that's all the difference a belief in Kevin makes to Jim's life, why should he care about justifying it? Whether the belief is really true or only the stuff of fables makes essentially no difference to how Jim chooses to spend his time, with whom Jim cares to associate, where Jim will live and work, whether Jim will retire at age sixty-five.

At the other extreme, suppose Jim is the leader of the local chapter of Kevinism. He wakes up every morning, drags his children from bed, and feeds them insects because if they're good enough for Kevin, they're good enough for Kevin's followers. Jim homeschools his children, immersing them in Kevin lore and doctrine, has them practice their croaking for half an hour, and then, after lunch, hops to the Kevin temple down the street, where he prepares the evening sermon. Moreover, because Jim is confident that Kevin would disapprove of camouflage-colored clothing, he spends his free time lobbying politicians to make illegal the sale and distribution of camouflage. Jim has also read in *The Life and Times of Kevin the Karnatakan Frog* that Kevin frowns on mixed marriages between vegetarians and meat eaters. Thus, Jim leads rallies on the Capitol steps, urging lawmakers to forbid such ungodly unions. Jim also pushes for legislation that would require all pregnant women to receive an exam to determine whether the fetuses growing inside them have extraordinarily long tongues. The procedure is very invasive, but Jim believes that the next holy leader of the Church of Kevin will be born with a lengthy tongue, and it's important to prepare for the event.

If this is what a belief in Kevin means to Jim, don't you think he should want to spend some time verifying it? Shouldn't a belief with this kind of significance for Jim's life *necessitate* justification? Imagine Jim's horror, if, near the end of his life, someone proves to everyone's satisfaction that Kevin in fact never existed. No doubt, Jim would want to resist this conclusion. After all, he's dedicated his entire life to worshipping and honoring Kevin. But suppose that the evidence presented to Jim is so completely compelling that even he must concede that he was wrong. There was never any Kevin. Jim should be ashamed of himself. He allowed a belief for which he had inadequate justification become absolutely central to his life. Worse, he tried to regulate the lives of others on the basis of this belief.

I have described two cases that sit on either end of a spectrum. As I see it, on the one hand, when a belief hardly matters to you—when it's of minimal significance—it's not one you need to justify. Why bother? On the other hand, as the belief increases in significance, the motivation for seeking justification for it should also rise. Ultimately, as in the case of the second Jim, justification for the belief becomes hugely important, and failure to seek justification becomes something like a moral lapse. Why a moral lapse? Because not only does Jim fail himself—deprives not only himself of a life that could have been lived more meaningfully—but he also seeks to deprive his children and others of the autonomy to eat what they want, wear want they want, marry whom they want, and make decisions about their own bodies.

So should it matter to you whether you lack justification for your beliefs in miracles? That depends. How significant in your life are these beliefs? Would you live your life differently if you didn't believe in miracles? How differently? Maybe you derive value from a sense of belonging to a religious community, and that's your primary reason for attending religious functions, observing religious holidays (holy days), and so on. You believe in miracles, but this belief is mainly incidental to how you live your life. Were someone to convince you that miracles were impossible, you would express some surprise but continue living your life pretty much as you had before. If I am describing you, then I would say that the arguments I have made in the preceding chapters should come as little disappointment. Study them for their intellectual interest.

But maybe belief in miracles is more significant for you. Maybe you're closer to the second Jim. Your belief in miracles dictates a particular life for you. Because you believe, you keep to a special diet despite wishing you could eat a particular kind of food—for example, bacon. You spend a great deal of money to send your children to a private religious school—money you might have used otherwise to help the poor and starving or, more selfishly, to vacation in the Bahamas. You vote for politicians who would forbid abortion or who work to prevent gay marriages. If this is who you are, I would say the arguments I have made in this book should be *very* troubling. You're at fault if you do not seek to justify those beliefs that figure so prominently in how you choose to live your life and, especially, how you choose to constrain the lives of others.

In conclusion, I can tell you that belief in miracles is unjustified. But only you can say whether that matters to you.

APPENDIX I

WHAT IS SUPERNATURAL?

A TEMPTING THOUGHT IS THAT AN EVENT IS SUPERNATURAL IF its occurrence violates a law of nature. This seems to be what's going on with brooms that fly or frogs that heal. We can describe laws of nature—for example, the law of gravity or laws about physiological processes that prevent human limbs from regenerating—that ought to make these sorts of things impossible. So when such things occur, the laws of nature are being broken. But what are these laws of nature that are purportedly being violated? Whatever they are, they are not like the sorts of laws that governments impose on citizens or that clubs place on their members. Suppose you are visiting a country and are not sure whether you are old enough, according to its laws, to drink alcoholic beverages. You can look up the drinking-age law and find out whether you can legally drink. The law is the product of human decisions and might change over time and might differ from place to place.

Laws of nature are not like this. Newton didn't simply make up the law of gravity. It's not true because people decided that it was for some reason necessary to keep order in society. He inferred it from observations of falling and rolling objects. Gravity, to use terminology we introduced earlier, was Newton's inference to the best explanation of falling apples, tides, orbiting planets, and so on. But if laws of nature are discovered by observing how things in the world work, how can anything ever violate a law of nature?

To see the problem, let's suppose it is a law of nature that water freezes at 32°F. We believe that this is a law because we have many observations of water freezing at just this temperature but no observations of water freezing at some other temperature (barring things such as changes in altitude, impurities in the water, and so on). How might this law be violated? You come across a sample of water, perhaps in a puddle in your driveway following last night's rainstorm, that fails to freeze when you place a sample of it in your freezer. It's a law that water freezes at 32°F, but this water does not, so we have a clear case of a law being violated. Presumably this is how to understand miracles as well. It's a law that bodies denser than water should sink, but the Bible says that Jesus's body, which was denser than water, did not, so when Jesus walked on water, he violated a law of nature.

But remember: laws of nature are not like vagrancy laws or other laws that come about through decisions made by groups of individuals (or perhaps by single individuals in autocratic societies). We believe that the law about the freezing temperature of water because of the many observations that we can cite in its support. When you place a sample of water in your freezer and it refuses to freeze, we now have a new observation, different from all our earlier observations. What should we say about the law? Has it been violated?

We seem to face a choice. On the one hand, if we wish to stick to our guns and insist that the law really is a law, that it's really true that all water freezes at 32°F, then we have to deny that the stuff you have put in your freezer is water. It can't be water because it doesn't freeze at 32°F. In this case, the law hasn't been violated because the law says nothing about how substances other than water behave. It describes only how water behaves, so if the law is true, the substance that you removed from the puddle on your driveway can't be water.

On the other hand, we might say that we were wrong when we thought that all water freezes at 32°F. Remember—laws of nature are not simply written down somewhere waiting for us to discover them. We infer them from our observations. Now we have an observation of water that doesn't freeze at 32°F. That means that we have to go back to the drawing board, as it were, and reformulate the law to accommodate this new observation. Perhaps the correct law is that water freezes at 32°F *unless it comes from puddles following rainstorms*. I agree that this sounds like a strange law, but it does fit the observations. And when you think about other statements we accept as laws, maybe it's not so strange after all. Imagine explaining to someone who has grown up in a tropical climate and has never heard of snow that rain is a liquid until the temperature becomes very cold, after which rain becomes tiny white crystals that stick to the ground and can be shaped into balls and thrown at cars. That's pretty weird too.

In any event, the point is that once we have reformulated the law of nature to take into account our latest observation, we again are left without a violation. According to our new law, the stuff in your freezer is behaving exactly as it should because water from puddles does not freeze at the same temperature as water in ponds or from faucets or in fish tanks.

Either way, then, what we might initially have thought to be a violation of a law of nature turns out not to be. In the first case, we accept that all water freezes at 32°F, and when we come across something that freezes at some other temperature, we simply deny that it is water. The law stays true. Or we modify our law so that apparent exceptions to the law no longer count as exceptions, as when we stipulate that not all water must freeze at 32°F. Again, no violation.

Turning now to miracles, it's easy to see that we face the same difficulties when we try to understand their supernaturalness in terms of violations of natural laws. Suppose Jesus did walk on water but also that it's a law that objects denser than water sink. Does Jesus's action violate a law of nature? Our options are the same as they were in the case of the water that apparently fails to freeze. Either it really is a law that objects denser than water sink and the law was not violated because Jesus's body must not have been denser than water, or Jesus's body was denser than water, but we were wrong about the law. The law, as revealed by further observations, must be something more like the following: objects denser than water sink unless the objects are Nazarene carpenters. Similarly, either it's a law that frogs can't heal amputees and the Karnatakan frog is not really a frog after all, or, somewhat surprisingly, the real law is that frogs can't heal amputees unless they are from Karnataka.

So what is the believer in miracles to do? How ought we to understand the supernaturalness of miracles if not in terms of violations of natural laws? Here's an idea that might help. If we're to believe in miracles, we must accept, in effect, the existence of two kinds of worlds. There's the natural world, full of things such as water, rocks, gravity, mass, and so on. This is the world governed by laws of nature. Then there's the supernatural world, full of things such as angels, God or gods, ghosts, witches, and so on. Of course, I think we should agree that we have a much better sense of what exists in the natural world than in the supernatural one, but to make sense of miracles we should simply grant the existence of a supernatural world even if we're unsure exactly what it contains. What happens now when a miracle takes place? Suppose, for instance, a marble statue of Mary weeps genuine tears. This is impossible according to the laws of nature. However, it's not impossible given supernatural interference in these laws. Due to the influence of some supernatural agent, the statue begins to cry. But this doesn't count as a violation of a law of nature because laws of nature describe only events in the natural world. Supernatural causes are outside this natural world, and though they can (by assumption) influence goings on in the natural world, their doing so doesn't require that we reformulate the laws of nature. When the statue cries, the laws of nature that prevent such a thing are in effect being overridden.

An analogy might help to make this idea clearer. Suppose that you were to place a steel ball, like a pinball, on top of an inclined piece of plywood. When you release it, it rolls down the plane in a path determined by the laws of gravity and friction. However, if we wanted to, we could interfere with the path that laws of nature normally dictate. We could screw pegs into the board that divert the ball from the path that it would have followed, causing it to follow instead a different path. These interferences are analogous to the supernatural causes that produce miracles. The laws of nature that describe the path the ball would have taken were it not for the pegs are still true, but they can't be "blamed" for failing to take into account the influence of the pegs because, in a sense, that's not their business.

Now, when a statue of Mary cries, we should understand this miraculous event not as a violation of a law of nature, but as a supernatural interference with this law. The laws of nature continue to be true, just as the law of gravity continues to hold when the steel ball rolls down the inclined plane. It's just that something—a supernatural force—now interferes with the laws that hold true of marble statues, causing the statue to behave in a way that it normally would not. The statue cries as a result of this interference.

So I think we can say that a coherent interpretation of the supernaturalness of miracles is possible (barring worries I raise in appendix 2). It's an interpretation that says, in effect, that the natural world behaves in a regular way—a way that laws of nature describe—unless something supernatural jumps into the picture, interfering with the normal course of events. So what does it mean to say that some event is supernatural? It means that the normal course of nature has been interfered with by something outside of nature, so that we can no longer count on the laws of nature to produce their normal results. Thus, marble statues may cry, human bodies may walk upon water, and frogs may resurrect pets (unless the pets are ferrets).

APPENDIX 2

SUPERNATURAL CAUSES

LET'S TALK ABOUT POOL. NOT SWIMMING POOLS, BUT THE GAME that's played on a green felt surface with solid and striped balls. You use a cue stick to hit a cue ball, which, ideally, strikes another ball, and, if all goes according to plan, your desired target ends up in one pocket or another. I trust that everyone's familiar enough with the game to know what I'm talking about. Let's now imagine a slightly different game. As before, you hold a cue stick in your hands, but in this game there's no cue ball. You still have to hit the cue ball with the stick to move the other balls on the table. How do you do this without a cue ball?

Instead of a real cue ball, this game is played with an imaginary cue ball. What you do is this. You imagine a cue ball at a particular spot on the table. You then line up your shot, calculating at what angle you need to hit the imaginary cue ball with your cue stick so that it will smack into, say, the four ball, which, you hope, will end up in the corner pocket.

I invite you to try this game next time you find yourself playing pool with some friends (to save yourself some embarrassment, you might want to try it on your own first). I'll bet you anything that not a single ball on the table moves as a result of your hitting the imaginary cue ball. The four ball will stay exactly where it was even if you hit the imaginary ball at just the angle you wanted to.

Saying why a game of pool with an imaginary cue ball might be frustrating is pretty easy. The imaginary cue ball can't actually cause any of the other balls on the table to move because it has no mass. Without mass, it can't possibly exert a force on anything and so can't move anything. Without mass, it has no energy to transfer to the four ball; thus, the four ball stays put when "hit" with the imaginary cue ball.

Let's now replace the imaginary cue ball with a supernatural one. Or, in case there are no supernatural cue balls, let's suppose that a supernatural force of some kind wants to cause the four ball to move. Perhaps an angel has put some money on your game and realizes that you have made a bad shot. He's now going to correct the trajectory of the four ball by pushing it with his invisible hand. He extends his supernatural arm and gives the ball a flick with his supernatural index finger. What happens?

The reason the imaginary cue ball couldn't move the four ball was that it lacked mass and so had no energy to transfer. But what about the angel's finger? Supernatural entities are, by definition, outside nature. That means that they have no natural properties. They don't have mass. They don't have volume. It makes no sense to ask how long an angel's finger is or how much it weighs. The common idea that angels are sort of like ghosts or that ghosts are sort of like people that you can see through is incoherent. How tall is an angel? How tall is God? How much does a ghost weigh? If angels, ghosts, God, and so on really are supernatural, then they don't have shapes at all; they don't weigh anything. They don't walk through walls because they don't walk because they don't have legs. They don't fly through the clouds because they don't have wings. Paintings of winged angels shouldn't be interpreted literally. At best, the artists are taking liberties with the idea of supernatural beings by presenting them to us in the only way possible—by making visible something that, strictly speaking, cannot ever be visible, that has no shape, that reflects no light.

Angels, if really outside of nature, don't have fingers and can't extend their arms toward moving cue balls. And even if they did, and even if they could, I don't see how that would help your pool game. How could an angel move a cue ball if it can't touch the ball? It could no more cause the ball to move than could the imaginary cue ball.

As I said at the beginning of this discussion, I don't know of any satisfactory solutions to this problem: the problem of how supernatural entities—things that exist outside of nature—can interact with things within nature. If anyone says that they do, don't believe them. The most anyone can say is just this: it happens, don't ask how. As a critically thinking philosopher, I can't condone this answer. However, I can let it slide in the present context so that we're free to examine whether belief in miracles is justified.

NOTES

Preface

1. Pew Forum on Religion & Public Life, *Religion Among the Millennials* (Washington, D.C.: Pew Research Center, February 2010), http://www.pewforum. org/files/2010/02/millennials-report.pdf.

2. Miracles

1. See Baruch Spinoza, *Theological–Political Treatise* (1670), 2nd ed., trans. Samuel Shirley (Indianapolis: Hackett, 2001).

3. Justifying Belief in Supernatural Causes

1. John Locke, *The Reasonableness of Christianity, with A Discourse of Miracles and Part of a Third Letter Concerning Toleration*, ed. I. T. Ramsey (Stanford, Calif.: Stanford University Press, 1958), 78–87.

2. Ibid.
3. Bart Ehrman, *Misquoting Jesus* (New York: HarperCollins, 2005), and *Jesus, Interrupted* (New York: HarperCollins, 2009).
4. Michael Licona, *The Resurrection of Jesus: A New Historiographical Approach* (Downs Grove, Ill.: InterVarsity Press, 2010), 177.
5. Ibid.

4. Justifying Belief in Improbable Events

1. Angus Reid Public Opinion, *A Vision Critical Practice* (2012), http://angusreid global.com/wp-content/uploads/2012/03/2012.03.04_Myths.pdf.
2. David Hume, "Section X," in *An Enquiry Concerning Human Understanding* (1777), 2nd ed., ed. Tom Beauchamp (Oxford: Clarendon Press, 2006), 83–89.
3. Michael Licona, *The Resurrection of Jesus: A New Historiographical Approach* (Downs Grove, Ill.: InterVarsity Press, 2010), 144–145.
4. Norman Geisler, "Miracles and the Modern World," in *In Defense of Miracles: A Comprehensive Case for God's Action in History*, ed. R. Douglas Geivett and Gary R. Habermas (Downer's Grove, Ill.: InterVarsity Press, 1997), 79.
5. C. S. Lewis, *Miracles* (New York: Macmillan), 102.
6. Winfried Corduan, "Recognizing a Miracle," in Geivett and Habermas, *In Defense of Miracles*, 109.
7. Ibid.
8. Licona, *The Resurrection of Jesus*, 175.

5. Evidence for Miracles

1. Richard Carrier, *Why I Don't Buy the Resurrection Story*, 6th ed. (Secular Web, 2006), http://infidels.org/library/modern/richard_carrier/resurrection /lecture.html.
2. Douglas Geivett, "The Evidential Value of Miracles," in *In Defense of Miracles: A Comprehensive Case for God's Action in History*, ed. R. Douglas Geivett and Gary R. Habermas (Downers Grove, Ill.: InterVarsity Press, 1997), 186.
3. Michael Coe, "Mormons and Archaeology: An Outside View," *Dialogue: A Journal of Mormon Thought* 8, no. 2 (1973): 42.

6. Jesus's Resurrection

1. Bart Ehrman, *Misquoting Jesus* (New York: HarperCollins, 2005), and *Jesus, Interrupted* (New York: HarperCollins, 2009); Richard Carrier, "Why I Don't Buy the Resurrection Story," in *Internet Infidels, Inc.* (2006), http://www.infidels .org/library/modern/richard_carrier/resurrection/lecture.html.

2. Wendy Cotter, *Miracles in Greco-Roman Antiquity* (New York: Routledge, 1997).

3. N. T. Wright, *The Resurrection of the Son of God*, vol. 3 of *Christian Origins and the Question of God* (Minneapolis: Fortress Press, 2003), 710.

4. Ibid., 717.

5. Ibid., 687.

FURTHER READING

AS YOU MIGHT EXPECT, A NEARLY LIMITLESS AMOUNT OF literature has been written on the topics of justification, miracles, and the gospels. In writing this book, I read quite a bit of this literature but still only scratched the surface. In addition to the material cited in the notes, I can recommend the following works as especially valuable.

On the Nature of Miracles

Corner, David. "Miracles." In *The Internet Encyclopedia of Philosophy*, edited by James Fieser and Bradley Dowden. 2009. http://www.iep.utm.edu/miracles/.

Houston, J. *Reported Miracles: A Critique of Hume*. Cambridge: Cambridge University Press, 1994.

McGrew, Timothy. "Miracles." In *The Stanford Encyclopedia of Philosophy*, edited by Edward Zalta. Stanford, Calif.: Stanford University Press, 2011. http://plato.stanford.edu/entries/miracles/.

On Arguments Concerning Justification for Belief in Miracles

Earman, John. *Hume's Abject Failure: The Argument Against Miracles*. New York: Oxford University Press, 2000.

Fogelin, Robert. *A Defense of Hume on Miracles*. Princeton: Princeton University Press, 2003.

Millican, Peter. " 'Hume's Theorem' Concerning Miracles." *Philosophical Quarterly* 43 (1993): 489–495.

Schlesinger, George. "Miracles and Probability." *Noûs* 21 (1987): 219–232.

Sobel, Jordan. "On the Evidence of Testimony for Miracles: A Bayesian Interpretation of David Hume's Analysis." *Philosophical Quarterly* 37 (1987): 166–186.

On the Historicity of the Gospels

Blomberg, Craig L. *The Historical Reliability of the Gospels*. Downs Grove, Ill.: InterVarsity Press, 2007.

On the Greco-Roman Context of the Gospels

Jeffers, James. *The Greco-Roman World of the New Testament Era: Exploring the Background of Early Christianity*. Downs Grove, Ill.: InterVarsity Press, 1999.

Levine, Amy-Jill, Dale Allison, John Crossan, eds. *The Historical Jesus in Context*. Princeton: Princeton University Press, 2006.

INDEX

Aaron, 26, 52, 54–55, 58, 109, 137,
  139–141; and the serpents,
  41–49
Aesclepius, 121
Akita. *See* Our Lady of Akita
alien abduction, 69–73; 88, 103, 112,
  118, 133–135. *See also* extraterrestrials;
  Sally (example)
Angus Reid Public Opinion poll,
  74
Apollonius, 121–123
apologists, 91, 142
Aristotle, 10–11
Athena, 93
Augustus (emperor), 130–131

Babel, 99, 105
background assumptions, 37–40;
  140–141; and Aaron's staff, 42–44, 46;
  and the supernatural, 48–52, 55–56;
  58. *See also* inference to the best
  explanation
base-rate fallacy, 64–66, 68
Bayes, Thomas, 85
Bayes's theorem, 85
Beatles, 61–62
begging the question fallacy, 82, 131
belief: justified and true vs. unjustified
  and true, 7–9; justified vs. unjustified,
  6–18, 138–139; true vs. false, 6; vs.
  wishes, 4, 17–18

Bible, 91, 114, 124–126, 150. *See also*
    gospels; New Testament
Bigfoot, 70, 73–77, 132. *See also* Sasquatch
Blaine, David, 47

Caesar, Julius, 52, 90–96, 99, 102–103,
    105–106, 112, 114, 116–117, 128–132, 134
Caligula (emperor), 52
Carrier, Richard, 91–93, 95–97, 113,
    160–161
Catholics, 50, 125
Christ, 98, 101. *See also* Jesus
Christianity, 115, 124, 128–132, 159, 164
Church of Jesus Christ of Latter-day
    Saints, 98, 101. *See also* Mormonism
Church of Kevin (example), 147
Coe, Michael, 102
Corduan, Winfried, 83

Damascus (pope), 125
Darwin, Charles, 10

Ehrman, Bart, 53, 113, 124
Einstein, Albert, 11
Elijah, 121–123
Elisha, 121–123
Empedocles, 121, 123
epistemology, 5
Erasmus, Desiderius, 125–126
Euclid, 5
evangelicals, 103
evangelism, 114, 126
evidence: for Bigfoot, 74–77; five kinds of
    in historical research, 93–97; historical,
    for miracles, 91–92, 109; improbability
    of, for a miracle, 21–22, 24–25; from
    independent sources, 79–81; for Jesus's

resurrection, 52–54, 110, 112–119,
    129–130; and justification, 11, 14, 18;
    for Mormonism, 97–107; for the
    supernatural, 57–58; type necessary
    for belief in miracles, 85–86, 88–90,
    134–135, 141–142. *See also* justification
Exodus, 41
explanation, 7–9, 15. *See also* inference
    to the best explanation
extraterrestrials, 24, 49–51, 56, 58, 72,
    88, 140–141

faith, xv, 50, 72, 116, 137, 141
Flavius Josephus, 115
frog, 2–5, 14–15, 19–21, 24–25, 30,
    59–60, 83, 87–90, 92, 110–111, 118,
    137–138, 140–143, 146–147, 152. *See
    also* Kevin (Karnatakan frog example)

gambler's fallacy, 6
Geisler, Norman, 78–81, 85
Geivett, Douglas, 91–92, 97
Genghis Khan, 97
ghosts, xiv, 25, 30, 144, 152, 156
gods, 46–48, 54–55, 121, 152
Goldbach's conjecture, 6
gospels, 52, 91, 111–119, 121–129, 134–135

Herod Antipas, 114
Holocaust, 13, 134
Hume, David, 78
Hyppolytus, 121

inference to the best explanation, 33–38, 83,
    85, 98, 139–140, 150; and Aaron's staff,
    40–49, 52; and Jesus's resurrection,
    53–55; and the supernatural, 55–58;

and weeping statues, 49–50, 52, 55–56,
58. *See also* justification
Irenaeus of Lyons, 114
Israelites, 41, 98

Jared, 99
Jaredites, 99, 101, 103
Jebbediah (example), 122–123
Jerome, 125
Jerusalem, 100, 116, 122, 130
Jesus, 60, 78; in Mormonism, 98–102,
105–106; and resurrection, 52–55, 79,
81, 83–84, 91–92, 97, 107, 109–140;
and the resurrection of Lazarus, 26;
and walking on water, 31, 150–151
Jim (example), 2–5, 14–15, 21, 24–25, 30,
59–60, 87–90, 92, 110–111, 113, 118,
137–143, 146–148
John, xiv, 111, 114, 116, 125 128
Josephus. *See* Flavius Josephus
Julius Caesar. *See* Caesar, Julius
justification: for belief in Jesus's
resurrection, 112–135; for belief in
Mormonism, 97–107; vs. explanation,
6–8, 15–16; vs. faith, 137, 141; for
historical beliefs, 93–97; and knowledge,
6; objectivity of, 12–13; as probability
raiser, 7, 9–10, 13–14, 18, 138; and the
supernatural, 29–32, 35, 56–58; and
testimony, 15–16, 63, 69, 72, 78, 85;
and truth, 4, 6, 9–12; vs. unjustified,
6, 9. *See also* belief; evidence; inference
to the best explanation

Kennedy, John F., 129, 134
Kevin (Karnatakan frog example), 110–111,
118, 138–143, 146–147. *See also* frog

Kevinism, 146
Khan, Genghis. *See* Genghis Khan
King James Bible, 125–126
knowledge, 5–6. *See also* epistemology

Lamanites, 100–101, 103
Lazarus, 26, 121, 127, 134
lepers, 23, 52, 121
Lewis, C. S., 82–85
Licona, Michael, 53–55, 78–81, 84–85
Lincoln, Abraham, 90, 130, 132
Loch Ness monster, 73, 76
Locke, John, 40–46, 48–50, 52
Luke, xiv, 111, 114, 116, 126–128

Magdalene, Mary, 127
Margaret (example), 35–36, 45
Mark, xiv, 111, 114, 116, 126 128
Mary, 31–32, 35, 49–51, 55–58, 61,
152–153
Matthew, xiv, 111, 114, 116, 126–128
miracle(s): of Aaron's staff, 40–48, 52,
55, 58; definition of, 18; historical
evidence for, 91–92, 96–97, 106;
improbability of, 20–25, 57, 59–61,
77–78, 85–86; Jesus's resurrection,
53–55, 84, 121; of Our Lady of Akita,
49–52, 55–56, 58; supernatural origin
of, 25–26, 30, 57. *See also* evidence;
inference to the best explanation
Miranda (example), 33–40, 42, 44–46,
48, 140
Mongols, 97, 99
Mormonism, 105–106, 110, 113
Mormons, 97–99, 100–105, 109–110
Moroni, 100, 102
Moses, 26, 41–43, 109, 122, 137, 139–140

Nancy (example), 33–34, 36–42,
  44–46, 48
Nephites, 100–101, 103
New Testament, 3, 52, 111–113, 124,
  130–131, 133
Nixon, Richard, 70–71, 94

Odysseus, 93
Old Testament, 123
Oracle of Delphi, 89
Our Lady of Akita, 49, 51–52, 55–56

Patty (example), 65–67
Paul, 115
Pew Forum on Religion survey, ix–x
Pharaoh, 41–44, 46–48
Pliny the Younger, 115
Plutarch, 94, 117
Pompeii, 81, 130
Pontius Pilate, 114–115
pustulitis (example), 63–72, 75–76, 81,
  92, 141–142

Rachel (example), 45–46
reason, x–xv, 7; vs. wishing, xii–xiii
resurrection: See evidence;
  inference to the best explanation;
  Jesus; miracle(s)
Roswell, N. M., 73
Rubicon River, evidence of Caesar
  crossing, 52, 90–96, 99, 102, 105–106,
  112, 114, 116–117, 128–132, 134

Sally (example), 68–73, 79, 88, 92, 103,
  112, 118, 132–135. See also alien
  abduction
Sasquatch, 8–9. See also Bigfoot

Semmelweis, Ignaz, 34–35
serpent, 40–48, 54, 58, 109, 137, 141.
  See also Aaron
Smith, Joseph, 100, 102–107,
  118, 132
Spinoza, Baruch, 18, 51, 144
Stanley the statistician (example), 64–67
stigmata, 31
supernatural, 41, 46–52, 60, 98, 139–141,
  149, 152–153; causes of the, 155–157;
  improbability as evidence for the,
  18–19, 21–22, 24, 26, 55; justifying
  belief in, 29–32, 35, 56–58. See also
  evidence; inference to the best
  explanation; justification; miracle(s)
synoptic gospels, 114

Tacitus, 115
testimony, 68, 71, 78–81, 86, 90, 96; and
  Bayesianism, 85; on behalf of Bigfoot,
  74–77; on behalf of miracles, 92,
  121–122, 135; as justification, 15–16,
  59, 61–63, 69, 72–73. See also
  evidence; justification
Thomas, 128
Thomson, J. J., 34–35
Titanic, 130
tomb, of Jesus, 53, 79, 115–116, 123, 127,
  130–131, 133–134
trinity, 125–126

Vespasian (emperor), 122
Virgin Mary. See Mary

Wisconsin, 22, 41
witches, 30, 93, 144, 152
Wright, N. T., 130–133